Economic Geography

This book series serves as a broad platform for scientific contributions in the field of Economic Geography and its sub-disciplines. *Economic Geography* wants to explore theoretical approaches and new perspectives and developments in the field of contemporary economic geography. The series welcomes proposals on the geography of economic systems and spaces, geographies of transnational investments and trade, globalization, urban economic geography, development geography, climate and environmental economic geography and other forms of spatial organization and distribution of economic activities or assets.

Some topics covered by the series are:

- Geography of innovation, knowledge and learning
- Geographies of retailing and consumption spaces
- Geographies of finance and money
- Neoliberal transformation, urban poverty and labor geography
- Value chain and global production networks
- Agro-food systems and food geographies
- Globalization, crisis and regional inequalities
- Regional growth and competitiveness
- Social and human capital, regional entrepreneurship
- Local and regional economic development, practice and policy
- New service economy and changing economic structures of metropolitan city regions
- Industrial clustering and agglomeration economies in manufacturing industry
- Geography of resources and goods
- Leisure and tourism geography

Publishing a broad portfolio of peer-reviewed scientific books *Economic Geography* contains research monographs, edited volumes, advanced and undergraduate level textbooks, as well as conference proceedings. The books can range from theoretical approaches to empirical studies and contain interdisciplinary approaches, case studies and best-practice assessments. Comparative studies between regions of all spatial scales are also welcome in this series. Economic Geography appeals to scientists, practitioners and students in the field.

If you are interested in contributing to this book series, please contact the Publisher.

More information about this series at http://www.springer.com/series/15653

Felipe Irarrázaval · Martín Arias-Loyola
Editors

Resource Peripheries in the Global Economy

Networks, Scales, and Places of Extraction

Editors
Felipe Irarrázaval
Instituto de Estudios Urbanos y Territoriales
Pontificia Universidad Catolica de Chile
Santiago, Chile

Martín Arias-Loyola
Departamento de Economía
Universidad Católica del Norte
Antofagasta, Chile

ISSN 2520-1417 ISSN 2520-1425 (electronic)
Economic Geography
ISBN 978-3-030-84608-4 ISBN 978-3-030-84606-0 (eBook)
https://doi.org/10.1007/978-3-030-84606-0

This Springer imprint is published by the registered company Springer Nature Switzerland AG
The registered company address is: Gewerbestrasse 11, 6330 Cham, Switzerland

Contents

Editors and Contributors

About the Editors

Felipe Irarrázaval is Postdoctoral Researcher and Adjunct Professor at the Instituto de Estudios Urbanos y Territoriales at the Pontificia Universidad Católica de Chile and at the Center for the Study of Conflicts and Social Cohesion (COES). He received a Ph.D. in Human Geography from the University of Manchester. His research examines resource governance, particularly extractive industries in Latin America, through the lens of global production networks, political geography, and urban studies. He has undertaken research in Peru, Bolivia, and Chile, and has published his research in journals like Economic Geography, Political Geography, Annals of AAG, Capitalism Nature and Socialism, and EURE.

Martín Arias-Loyola has received a Ph.D. in Economic Geography and Planning Studies from the University College London (UK) and M.Phil. in Regional Sciences from the Universidad Católica del Norte (Chile). He is Assistant Professor of Economic Geography at the Departamento de Economía and researcher at the IDEAR of the Universidad Católica del Norte in Antofagasta (Chile). He is also a Visiting Academic/Postdoctoral Researcher at the Faculty of Architecture, Building and Planning at the University of Melbourne (Australia). His research focuses on power asymmetries; the dark side of economic, political, and urban geography; extractivism; neoliberal extractive contexts; cooperativism; urban politics; mutual aid; informal settlements; and the role of academia in social change and the right to the city.

Contributors

Martín Arias-Loyola Departamento de Economía, Instituto de Economía Aplicada Regional (IDEAR), Universidad Católica del Norte, Antofagasta, Chile;

Faculty of Architecture, Building and Planning, University of Melbourne, Melbourne, Australia

Miguel Atienza Departamento de Economía, Instituto de Economía Aplicada Regional (IDEAR), Universidad Católica del Norte, Antofagasta, Chile

Beatriz Bustos Departamento de Geografía, Universidad de Chile, Santiago, Chile

Paul S. Ciccantell Department of Sociology, Western Michigan University, Kalamazoo, USA

Alexander Dodge Department of Geography, Norwegian University of Science and Technology, Trondheim, Norway

Maria Eugenia Giraudo School of Government and International Affairs, Durham University, Durham, UK

Felipe Irarrázaval Instituto de Estudios Urbanos y Territoriales, Pontificia Universidad Católica de Chile, Santiago, Chile

Juha Kotilainen Department of Geographical and Historical Studies, University of Eastern Finland, Joensuu, Finland

Julia Loginova School of Earth and Environmental Sciences, The University of Queensland, Brisbane, Australia

Kirsten Martinus School of Social Sciences, The University of Western Australia, Perth, Australia

María Inés Ramírez Departamento de Geografía, Universidad de Chile, Santiago, Chile

Marco Rudolf Universität Heidelberg, Heidelberg, Germany

Sören Scholvin Departamento de Economía, Universidad Católica del Norte, Antofagasta, Chile

Thomas Sigler School of Earth and Environmental Sciences, The University of Queensland, Brisbane, Australia

Michael John Watts Department of Geography, University of California, Berkeley, USA

Chapter 1
Introduction: Resource Peripheries in the Global Economy

Felipe Irarrázaval and Martín Arias-Loyola

Abstract The relation between resource extraction and the places in which extraction takes place has been a long-standing issue for academic, social and political debates. The paths through which resource extraction alter developmental dynamics, the everyday life of the local population and the environmental context have called the attention of social science since its origins. Despite the long-standing dimensions, which have been in the spotlight, contemporary political, economic and social changes demand revising the way in which resource extraction connects global production with the places where extraction occurs, here referred to as resource peripheries. This introduction critically revisits the academic debate about resource peripheries, asking to move forward from an understanding of resource peripheries as local models, towards a dynamic approach that allows grasping the socio-spatial relations that make the extraction places peripheral. For doing so, this section proposes three core dimensions that must be revisited in the research about resource peripheries: (i) changes in how contemporary capitalism is organizing production through globalized value chains; (ii) the re-scalation of political dynamics, which shape the economic organization of the places of extraction; and (iii) emergent issues, such as long-distance commuting, climate change and human rights

Keywords Resource peripheries · Production networks · Scales · Resource extraction

F. Irarrázaval (✉)
Instituto de Estudios Urbanos y Territoriales, Pontificia Universidad Católica de Chile, Santiago, Chile
e-mail: firarrazava@uc.cl

M. Arias-Loyola
Departamento de Economía, Instituto de Economía Aplicada Regional (IDEAR), Universidad Católica del Norte, Antofagasta, Chile

Faculty of Architecture, Building and Planning, University of Melbourne, Melbourne, Australia

M. Arias-Loyola
e-mail: marias@ucn.cl

F. Irarrazaval and M. Arias-Loyola (eds.), *Resource Peripheries in the Global Economy*, Economic Geography, https://doi.org/10.1007/978-3-030-84606-0_1

1.1 Introduction

The on-going way in which current capitalism is organizing the appropriation and transformation of nature into commodities is built over manifold and entangled socio-spatial relationships and the unsustainable imperative of endless economic growth. The elaboration of multi-scalar and socio-political regimes of capitalistic production, ensuring and facilitating resource extraction and the organization of complex production networks, are examples of the different socio-spatial processes underpinning the global race for the appropriation and exploitation of nature. Within such entangled and intricated relationships, the places where nature is extracted and transformed into economic resources are the starting point of the long trip that resources undertake through geographically and relationally embedded value chains. Global cities heavily rely on resource-rich territories to "fuel their own economies and lifestyles," (Hayter et al. 2003, p. 21). Yet, despite their paramount relevance for world's capitalism, resource-rich territories still remain relatively unexplored in comparison to the seas of ink poured into examining the metropolitan cores and secondary and/or tertiary industries.

The academic literature has studied resource-rich territories through different conceptual lenses, such as sites of extraction (Arboleda 2016), commodity source regions (Coe and Yeung 2015), extractive agglomerations (enclaves and clusters) (Arias et al. 2014; Phelps et al. 2015), resource curse (Cust and Poelhekke 2015), (neo)extractivism (Gudynas 2010; Svampa 2019) and resource peripheries (Hayter et al. 2003), to name a few. Even though those conceptualizations have deepened our understanding on the different processes happening at extraction sites, there is still a worrying lack of discussion regarding the positionality of the extractive places within the entangled socio-spatial relations shaping their incidental position in the global economic hierarchy and the potential dark sides of the geographies of extraction (Phelps et al. 2018). In this sense, despite the impressive growth rates in resource-based economies, the exploitation/extraction of natural resources is usually "a tide that tends to lift only some boats" (Argent 2017, p. 808) at the cost of extreme socioeconomic and environmental costs for extractive territories.

Moreover, recent processes have reconfigured how global capitalism interacts with resource peripheries, such as the organization of resource extraction at the global scale through complex, uneven and gendered value chains (Arboleda 2020; Baglioni and Campling 2017; Barrientos 2014; Gago and Mezzadra 2017), the emergent political dynamics from the localities in which extraction takes place forging multiscale alliances (Haarstad 2014; Riofrancos 2017), and how extractive urban nodes plugged into global extractive networks can act as backdoors to development, instead of gateways for greater value capture and territorial embeddedness (Atienza et al. 2020; Scholvin et al. 2019).

This book aims to reinvigorate the rich debate about resource peripheries that began almost two decades ago by uncovering how those places are produced through dynamic and integrated socio-spatial relationships. The classic contribution of Hayter et al. (2003) defined resource peripheries as marginalized places, starkly distinct from the cores, facing a specific set of processes or dimensions: industrialism (economic dimension), environmentalism (environmental dimension), aboriginalism (cultural dimension) and imperialism (geopolitical dimension). Ironically, even though they made a call to economic geographers to further addressing the ways in which the global economy impacts those places, resource peripheries remain at the fringe of this discipline's focus. Some authors have responded to Hayter et al.'s call (Argent 2013, 2017; Barton et al. 2008; Breul and Revilla Diez 2018; Ellem 2015; Hayter and Barnes 2012; McElroy 2018, among many others), but the debate still tends to focus on the processes taking place at resource peripheries instead of uncovering their causes.

Here, we propose to do exactly that, to shed light on the ways variegated socio-spatial relations underpinning the global economy are still (and probably will keep) producing resource peripheries with varying developmental outcomes so necessary for the cores' activities and subsistence. Some current conceptual developments that should be incorporated into this discussion include: the organization of production through gendered value chains (Bair and Werner 2011; Barrientos 2014; Coe et al. 2004), the disarticulation of actors from such globalized production (McGrath 2018), the reconfiguration of the scalar regimes of governance and power exertion in the last decades (Brenner 2004), and broader issues of ecosystemic (contra anthropocentric) and multi-scalar sustainable development (Svampa 2019), to name a few examples.

Likewise, current resource related tensions, such as climate change, political uprising against development strategies based on resource extraction and overall spatial inequalities keep multiplying the numbers of losers compared to the winners worldwide, meaning that academia should further contribute to the in-depth appraisal of the socio-spatial processes building and binding the fate of the places of extraction as peripheries to economic and political cores. As such, this book looks to stimulate the debate about resource peripheries by actively understanding them as a spatial category relationally produced within the global economy.

The contributions of this book aim to discuss the socio-spatial configuration of extractive places to better understand how they are socioeconomically and politically produced. They particularly scrutinize the positionality of resource peripheries within production networks and the multiple political scales that define their constituency as places of extraction and address the socio-spatial processes defining the sites in which the exploitation/extraction of resources takes place as peripheries. Before digging into the fascinating contributions of this book, this introduction briefly frames the debate about resource peripheries. Such framing not only states a critical revision regarding the almost twenty years of literature about resource peripheries. However, it also proposes three core processes in need of more debate regarding the socio-spatial production of the places of extraction as resource peripheries: (i) the industrial organization of resource extraction through globalized value chains; (ii) emergent scalar regimes that reframe the position of resource peripheries; and (iii) emergent issues that shape ongoing dynamics at the places of extraction, such as sustainability, commuting, and illicit actions.

1.2 Resource Peripheries: From Local Models Towards Socio-Spatial Relations

More than two decades ago, Hayter et al. (2003) called for a greater involvement of economic geography in the global exploitation and extraction of natural resources by claiming that "in the theoretical horizons of economic geographers, resource peripheries are a terra incognita," (p. 21). They considered that the theoretical approaches of the sub-discipline were mainly grounded in experiences related to the industrial cores and, consequently, they failed to become truly global and comprehensive perspectives. Their argument was that "globalization has different meanings, implications and history for resource peripheries than for cores," (Hayter et al. 2003, p. 18). Moreover, resource peripheries are relevant in terms of the scale of production and the unique experiences they host, both of which are crucial for further advancing economic geography's understanding of the world (Hayter et al. 2003).

Whereas the ongoing research on economic geography remains criticized due to its bias to successful cases in developed countries (Phelps et al. 2018), which usually excludes resource economies, it is worth acknowledging the rising number of economic geographers that study resource peripheries (Argent 2017; Barton et al. 2008; Breul and Revilla Diez 2018; Carson et al. 2016; Ellem 2015; Hayter and Barnes 2012; McElroy 2018), among many others. Likewise, political geography also plays an important role since territories are deliberately denominated as resource peripheries by political elites, governments and corporations (Argent 2017). This means resource peripheries are also an outcome of the power asymmetries between the resource rich territories and other actors located in the cores, instead of market-based processes. In this regard, and echoing Hayter et al. (2003), resource peripheries are a terra in which authors from different disciplines and approaches are exploring, uncovering the socio-spatial dynamics that take place in the areas of resource extraction.

The research about resource peripheries hitherto has been mainly concerned on the social, political, and economic processes taking place in a specific geographical scope, as a consequence of resource extraction. For instance, the classic contribution of Hayter (2003) points that a central characteristic of resource peripheries—as industrial regions—is the collision between neoliberalism, environmentalism and aboriginalism, which has produced a "local model" of institutions. As such, the purpose of Hayter (2003) and Hayter et al. (2003) is to note the active and dynamic processes of local contestation permeated by power imbalances that take place in regions located far from the cores, which is also where most resource exploitation and extraction occurs.

Likewise, Barton et al. (2008) looked to evolve the initial framework by discussing the possible transformations, rather than institutional dimensions, that are actively producing spatial heterogeneity at resource peripheries within a peripheral country. In this sense, Barton et al. (2008) proposed considering the dimensions by Hayter et al. (2003) as part of broader transformations: productive, environmental, sociocultural and political, where "each is a product of a set of institutional values implicit within the political economy and associated social structure that drives the contemporary development model," (Barton et al., p. 26). Such a development model is usually rooted in some form of accumulation strategy based on the appropriation of nature through large scale projects, especially in the Latin-American context. In this sense, classical extractivism has been defined as a development strategy, rooted in the colonial legacy of free-market exploitation of natural goods, with almost no value added, mostly for exportation, currently legitimized under a neoliberal logic. Conversely, neo-extractivism is also acknowledged as a development strategy, yet, fostered by left-wing states that took control of extractivism to promote and/or expand social policies aimed to poverty reduction (Gudynas 2010; Svampa 2019). However, as Hayter et al. (2003) warns, it is the extraction of natural resources with little processing to be sold elsewhere that is the root of the problems in resource peripheries and, consequently, to all the forms of accumulation strategies based in natural resources.

Hayter and Barnes (2012) advanced their institutional approach by discussing the geographical limits of the neoliberal expansion of resource extraction, building upon Anna Tsing's (2011) assemblage approach. Further contributions have discussed the effects of boom-and-bust cycles and climate change (McElroy 2018), demographic change (Carson et al. 2016; Nelson and MacKinnon 2004), infrastructure and urban development (Kirshner and Power 2015), labour dynamics (Ellem 2015), rent distribution (Argent 2013) or emergent forms of industrial organisation (Argent, 2017). Against this backdrop, and continuing the debate proposed by Hayter et al. (2003), economic geographers have advanced in making resource peripheries a terra 'more' cognita for the subdiscipline.

Despite the fruitful research progress about processes taking place within resource peripheries literature, the socio-spatial relations underpinning the definition of such peripheries have barely been discussed, especially against the ongoing organization of capitalist production through globalised production networks. Although noteworthy exceptions that systematically analyse the spatial functions of the places of extraction (Breul and Revilla Diez 2018; Steen and Underthun 2011) or regional analysis (Argent 2013; Carson et al. 2016; Ellem 2015), the literature about resource peripheries usually focuses on what happens within a specific scale (regions, cities or specific places), in which resources are exploited/extracted. This scale is frequently assumed as a periphery, due to its lack of industrial development related to resource extraction and physical distance from the—usually—metropolised cores. Hayter (2003) extended the industrial typology of Anna Markusen (1996), by defining resource peripheries as "slippery places" where resources and capital quickly flow away and where such flows are usually captured by a "sticky place" operating as a core.

Nonetheless, Hayter et al. (2003) also notice that incorporating theorisation and empirical analyses from outside the cores, like from the resource peripheries, is how economic geography can become a truly global discipline since they bring new, contextualised and more nuanced interpretations of the world. On this basis, Hayter (2003) looks to analyse industrial regions based on resource extraction through "local models" that synthesize the local and global dynamics of post-Fordism/globalised neoliberal forms of capitalism and, in particular, processes of local contestation against the neoliberal expansion of resource extraction (see also Hayter et al. 2003; Hayter and Barnes 2012). Particularly, Hayter et al. (2003) highlighted how resource peripheries have been starkly affected by post-Fordism, especially by its imperatives of flexibility and endless growth promoted by neoliberalism.

In this regard, the main purpose of the literature about resource peripheries has been to understand the places of extraction as local models, assuming their peripheral condition, rather than to unpack the socio-spatial relations defining their specialization systematically. As such, the concept of resource peripheries is a locational metaphor to address socio-political and environmental processes related to extractive industries, rather than to examine the interaction between different cores and the places of extraction. Acknowledging the positive influence that Hayter's approach on resource peripheries has had by unpacking the processes taking place at the sites of exploitation/extraction, the conceptualization of resource peripheries as local models eclipses the spatial dynamics and multi scalar arrangements underpinning the industrial organization of the resources industries. Beyond the pure theoretical interest in examining the socio-spatial relations underpinning resource peripheries, the following subsections address ongoing socio-spatial processes taking place in resource peripheries that calls to put forth more attention regarding their constituency as peripheral places: (i) networks, (ii) scales and (iii) emerging issues.

1.2.1 Changes in Networks

The rise of global value chains and production networks literature is heavily grounded in the spatial fragmentation of the productive process, in which value chains articulate more and more distant places that connect production and consumption centres in a topographical and topological manner (Allen 2016; Bair 2009; Coe et al. 2004; Gereffi 2018). Beyond the theoretical offshoot of these literatures, the empirical trend is the massive industrial reorganization that took place in different productive sectors during the last decades, which stands crucial for labour regimes and technological innovation (World Bank and WTO 2019).

This trend has been critical for resource industries in which leading industries orchestrate resource extraction through a diverse range of external services and suppliers located globally (Baglioni and Campling 2017; Bridge 2008). Whereas the research about production networks and resource extraction has critically pointed out that the contemporary organization of production under capitalism is reproducing the historical pattern of dependency, there have been changes regarding the involvement of peripheral actors at the production networks (Atienza et al. 2020; Breul and Rebilla Diez 2021; Scholvin 2020). The contemporary industrial organization of extractive industries operates through principles of flows, connectivity and speed that produce a distinctive configuration of production networks (Arboleda, 2020; Ciccantell and Smith 2009).

While three decades ago resource extraction involved the installation of massive settlements to appropriate nature and extract resources, such as company towns (Hayter 2000), nowadays, global industries plug into resource peripheries through gateway or backdoor cities (Atienza et al. 2020; Breul and Revilla Diez 2018; Scholvin 2020) and other forms of settlements, such as secondary cities, local settlements and extractive camps (Arias et al. 2014; Irarrázaval 2021; Phelps et al. 2015). This new entangled geography demands a more careful examination of the socio-spatial configuration of production networks from industrial cores towards the sites of extraction to grasp the contingent constituency of resource peripheries. Whereas some authors have proposed that this scenario offers attractive opportunities for industrial development at resource peripheries (Morris et al. 2012; Perez 2015), the agreement for the literature is that there have emerged new strategic cities at peripheral countries, but they have not broken down the historical threshold of technological and capital dependency that characterise resource extraction (Atienza et al. 2020; Breul and Revilla Diez 2021).

In any case, this new scenario is boosting new geographies of uneven development at the regional or country level, which asks for both a better comprehension of geographical form and the scalar logics that define resource peripheries. Werner (2019) and McGrath (2018) remind us that the dynamic processes involved in global production might lead to virtuous forms of strategic couplings, recouplings and even decouplings, but they could also force structural (re)couplings and the painful (dis)articulation of actors and territories. Beyond the acknowledged effects that extractive industries have on the livelihood of the population of resource peripheries due to their constitutive exclusion of production networks (Bebbington et al. 2008; Gudynas 2010; Svampa 2019), Werner (2016) states that global production networks boost uneven development through the new geographies of devaluation of the socially necessary labour time and processes of disinvestments at subnational level. In this regard, changes in production networks of resource extraction ask for a careful examination of the emergent geographies of devaluation and disinvestment that shape the places of extraction and other marginal scales.

1.2.2 Changes in Scales

Along with the changes on the industrial organisation during the last decades, there have been changes on the scalar configuration of political arrangements. Since the 90s, capitalist relations have leaned towards a spatial reorganization, where the institutional structures of the national state were aggressivelyrescaled, both upwards to global scales and downwards to local/regional scales (Brenner et al. 2003; MacLeod 2001; Swyngedouw 2004). Such a reconfiguration has reframed the political position of resource peripheries in different ways. From reconfiguration of the sub-national configuration of the states (Irarrazaval 2020) to cross-border alliance among states to facilitate infrastructure for extraction (Kanai 2016; Wilson 2011), resource peripheries are inserted into new scales of political arrangements that define their role at the international, spatialized and gendered division of labour (Gibson-Graham 1996; Massey 1984).

Such a reconfiguration is not only the result of global capitalism, but it is also shaped by local contestation at resource peripheries. Although the unrest of local communities is not necessarily new, political conditions during the last decades, such as the emergence of non-traditional forms of citizenship (Delamaza et al. 2017; Rich et al. 2019), decentralization reforms (Brosio and Jiménez 2012; Suarez-Cao et al. 2017) and international agreements (such as the ILO convention 169[1] or EITI[2]), allowed local communities to contest extraction in a more overt, empowered and constant manner, and re-scale their mobilization at different levels (Haarstad 2014; Riofrancos 2017). This scenario triggered a variety of multi-scalar political projects that looked to shape resource governance in multiple ways (Fry and Delgado 2018; Perreault 2018; Rasmussen and Lund 2018). For example, some local communities reject the materialization of extractive projects (Conde and Le Billon 2017), sub-national groups might challenge the distribution of the benefits related to resource extraction (Irarrazaval 2020) or political elites encouraging developmental projects based on resource extraction (Gudynas 2010; Svampa 2019).

Furthermore, issues related to the boom-and-bust cycles in the prices of natural resources have also been recognized as having peculiar effects over different scales (Hayter 2003). Yet, they have been overwhelmingly studied at the national scale. Extractive regions and urban spaces have historically shown clear distinctive effects during booms-and-busts over employment, infrastructure, housing, economic growth, and non-extractive sectors, spread unevenly among those spaces (Argent 2017). However, those effects remain veiled under methodological nationalism. Some booming cities, regions, and urban neighbourhoods can rip the benefits during periods of high commodity prices (Rehner et al. 2020), while others might turn into ghost territories, surrounded by assets abandoned during the bust phases, such as machinery and infrastructure (Argent 2017). Thus, as Rehner et al. (2020) stress, further empirical and conceptual studies about the multi-scalar socioeconomic effects of the extractive industries are crucial for "rescaling, spatializing and temporalizing" (p. 305) resource peripheries.

1.2.3 Emerging Issues

As the literature on research peripheries keep expanding, it must continue to incorporate the new and the old problems present in extractive industries. One, which is deeply related to the increasing fragmentation of production and spatial division of labour, is the fly-in fly-out (FIFO) of workers to and from the extractive sites. This

[1] The International Labour Organisation convention 169, introduced in 1989, is intended to protect indigenous communities. It has been ratified by 22 countries to date, including Bolivia and Peru.

[2] The Extractive Industries Transparency Initiative is a global standard for disclosing information about the contracts signed between firms and states.

has been noted as the "quintessential post-Fordist expression of flexibility" (Hayter 2003, p. 710), closely linked with rampant neoliberalism and labour flexibilization promoted by the cores in the Global North during the 70s and 80s (Arboleda 2020). Today, instead of building and maintaining costly extractive towns near the extraction sites, like during the early twentieth century, extractive firms have used the decrease in transport costs, especially in flights to import workers from distant regions for long working shifts, comprising of several days living in camps (Ellem 2015; Phelps et al. 2015). Officially, this has been done to reduce production costs, but as Ellem (2015) shows, the FIFO system has been weaponised against historically strong unions (such as in the mining industry), weakening their bargaining power and the territorial embeddedness between workers and extractive territories, while leaving resource peripheries extremely vulnerable to the negative externalities of extractive exploitation (Atienza et al. 2020).

The above is also linked with long-standing frictions between local grassroot actors organised to contest the worst consequences of extraction, such as extreme pollution, the destruction of aboriginal and cultural sites, the loss of biodiversity, genderised labour profiles in extractive industries, and overall damages to the environment and quality of life of human and non-human actors (Atienza et al. 2020; Svampa 2019; Vesco et al. 2020). This is permeated by power asymmetries, where local actors usually must face the power of international, multinational corporations and state-owned firms, defended by the ruthless use of state violence to supress activities affecting the normal workings of free-market neoliberal capitalism (Arias-Loyola 2021; Springer 2012). Thus, the challenges of climate change and human and nature rights will keep clashing with private and state-led violence and repression as long as the imperative of extractive-based growth keeps unfolding under neoliberal regimes (Temper et al. 2020; Bridge 2020). This raises issues of extractive governance and violence, especially in extractive contexts where power bases were established during colonial times (Barton et al. 2008; Bebbington et al. 2018; Springer 2012).

Even though those frictions are not new, they are reaching new topological and topographical scopes. As workers and civil society within extractive territories keep organising resistance to the accumulation by dispossession and repression promoted by the extractive global production networks (Argent 2017), the topological and topographical scale of this resistance keeps widening due to the dynamic creation and adoption of communication technologies. Such practices are promoting epistemological shifts, where local places can become central in the construction of "alternative global imaginaries," (Stephansen 2013, p. 506). Consequently, not only global trends against workers, aboriginal population and territories can start and be exported from local scales, such as the disarticulation of the powerful mining unions in Pilbara during the 1970s (Ellem 2015; Barrat 2019), but also structural changes aimed to overcome (neo)extractivist developmental strategies can be promoted from resources peripheries, such as international food sovereignty movements (Alonso-Fradejas et al. 2015).

In this regard, resource extraction will be continuously permeated by global challenges, such as climate change and sustainable development, which will reconfigure infrastructural design, waste management, and energy consumption. In any case, there is still much to be done, where issues regarding the automation/robotization of labour, new pandemics, the exploitation of renewable energies and biodiversity loss at extractive sites must be considered within the resource peripheries literature in the following decades. Whereas much of these processes are not necessarily new for resource peripheries, the literature has not addressed them in too much depth.

1.3 Examining the Socio-Spatial Dynamics of Resource Peripheries

The contributions of this book seek to advance the understanding of resource peripheries regarding both contemporary political and economic dynamics. Coherently with the ongoing socio-spatial processes taking places in resource peripheries described above, the book is organised in three sections: networks, scales and emergent issues. Even though this organisation is a methodological and practical exercise, the invitation is to read them interdependently and in an articulated fashion. The mentioned socio-spatial processes converge in resource peripheries, and the only way to properly examine, frame and understand the socio-spatial dynamics occurring at the places of extraction is through multi-dimensional perspectives capable of grasping as much as possible of what is happening there.

The networks section examines critical processes that are both reconfiguring the socio-spatial relations that define the places of extraction as peripheries and critically shape the ongoing developmental trajectories of resource peripheries. The contributions of this section share a value chains or production networks perspective, although from different traditions, to examine the ongoing way in which capitalism is organising the appropriation and transformation of nature into commodities. Paul Ciccantel's chapter undertakes the ambitious project of examining the shift of the socio-spatial relations that define coal peripheries at the global scale from a raw materialist lengthened global commodity chains model. Such an approach shares many premises of critical resource studies, such as the interest in examining the historical way through which socio-ecological relations shape the global race for resource extraction, with particular concern regarding the changes of transport and communications technologies that link manifold nodes of the commodity chains of raw materials. The historical analysis of coal extraction deployed by Ciccantel particularly informs the research agenda from a World System tradition that extends value chain analysis noting that "in a very material sense, transport systems are the sinews of hegemonic rivalry and stealing extractive peripheries."

Whereas the chapters of Alexander Dodge and Soren Schölvin also share the value chain perspective, they undertake a more conventional global production networks analysis to examine socio-spatial relations in resource peripheries. Schölvin particularly examines how the contemporary socio-spatial organization of hydrocarbon industries in Bolivia is downgrading the—once relatively developed—local industry, reinforcing poor developmental outcomes for purposes of industrial organization. This chapter shows the particular contradictions related to the apparent opportunities that the externalization of services by global industries offer to local firms and suppliers. Whereas global production networks offer chances to local actors, there are high entry barriers for local firms regarding their financial security. Consequently, new foreign companies enter the production network and compose a new link, in which they subcontract local suppliers and squeeze the latter profits. The deep production network approach of Schölvin allows us to unpack the way in which the highly fragmented form in which capitalism is organizing the exploitation of nature downgrade and excluding local industry.

Alexander Dodge provides an innovative debate for resource peripheries, particularly for those related to energy generation in Myanmar and Indonesia. His chapters mix global production networks and state analysis to examine the changing positionalities of production networks from merely exporters to emerging markets. Through a comparative analysis, he examines the socio-spatial processes that shape how resource peripheries, particularly those related to energy generation, control their resources to fuel local demand. In this regard, strategic selectivity based on developmentalist discourses are distinctive from the discourses related to the consolidation of energy markets in core countries, particularly in the context of the global market of liquefied natural gas. The chapter informs about the challenges and contradictions related to switching roles for the resource peripheries involved in production networks from exporters towards consumers, even of their own resources.

The contributions related to the scale section seek to uncover recent geographical reconfigurations of political arrangements that crucially shape the trajectories of resource peripheries. The chapter of Kirsten Martinus, Julia Loginova and Thomas Siegler discusses methodological and conceptual issues related to the places of extraction and proposes that "resource 'peripheries' are only peripheral insofar as they are politically marginalised or socio-economically disadvantaged." In this regard, they propose that peripheries should be understood relationally against a larger scale, in which they highlight three different scalar pairs that allows us to define different scales as peripheral: regional (non-metropolitan areas) and subnational; subnational and national; and national and global. Whereas resource peripheries share some commonalities, such as remoteness and banks of raw materials, they also vary depending on the specific socio-spatial processes that define their peripherality. The authors note that resource peripheries are both regional models to understand context-specific dynamics outside different cores and an outcome of the processes of uneven development triggered by resource extraction. In this regard, a crucial methodological strategy to properly understand the socio-spatial constituency of specific resource peripheries is a deep cross scalar examination, as they show for the Pilbara (Australia) and Northern Komi (Russia).

Juha Kotilainen's chapter provides an insightful conceptual and empirical approach to examine recent political arrangements that reconfigure the scalar constituency of resource peripheries by examining the raw materials policy of the European Union. He points that the global scope of extractive capital is limited by national states, which is something well know, but he extends the analysis upwards and downwards to discuss the role of regional blocks like the European Union and subnational politics in different cases. As such, the chapter discusses the role of non-state institutions and scales in defining resource peripheries because "the regional policies are not necessarily derived from the national ones in the sense that they would be governed from the nation-state governmental organs; these regional minerals policies have been much more independent." Kotilainen discusses how the concerns of European Commission regarding the imbalance between importation and exportation of resources lead towards a new spatial division among those who extract and use minerals within Europe, which is critically embedded in multi scalar discourses of extractivism and resource nationalism.

Maria Giraudo's chapter also uncovers scalar arrangements that are grounded in the state but exceed national scale from upwards and downwards, not only involving resource peripheries but also the body as a very situated scale. Through the scrutiny of the soy complex in Argentina, Brazil and Paraguay, Giraudo's claim that the particular scalar form of such a soy complex "scales is an indicator of a larger process of rescaling within the global political economy, whereby new spaces and temporalities for the organization of capital accumulation emerge and challenge the primacy of those already established." For the purpose of scalar analysis, Maria Giraudo's chapter is particularly innovative regarding the polymorphic socio-spatial relations that shape the soy complex, not only uncovering the global cores and the role of states but also explaining how the contingent scalar constituency of the soy complex penetrates into local livelihoods by altering the bodies of the local populations through toxic substances that flow through the scalar assemblage.

The chapters of the emergent issues section deal with some recent dynamics that take place in resource peripheries that are changing their interaction with cores of resource extraction. It is worth mentioning that some of those issues, such as the illicit dynamics examined by Michael Watt's chapter, are not relatively recent, but we consider that the literature hitherto has not dealt with them properly. Watts's chapter provides a detailed analysis of the production, intersection and hyper-complex features related to scale, specifically by empirically unravelling the political economy of global oil and gas assemblages. By fully embracing such daunting quest, Watts explores how the plethora of actors entangled in multi-scalar arrangements participate in both formal and illicit supply chains, providing a rich depiction of usually under-researched topics within the literature on resources peripheries related to the relation between licit and illicit production. Hence, by focusing on how power is exerted and value is produced and captured along these mixed relations within global value chains, he skilfully discusses empirical evidence about corruption, physical violence, and piracy co-existing with traditional state and market institutions throughout different nations and territories. The chapter greatly contributes to research peripheries literature by stressing how resource peripheries are dynamically and socially co-produced by several actors acting in a dis/organized manner, highlighting how both illegal and legal practices overlap to create different scales of extractive exploitation.

Studying the environmental and socioeconomic impacts that waste can have over an extractive territory and the usefulness of implementing a Circular Economy model,

Bustos, Ramírez and Rudolf provide an empirically rich analysis of the salmon industry in Chile. In this chapter, the authors reflect on the possibilities of implementing such a model that would entail breaking with the usual linearity of extractive-based processes, where waste is disposed of in the same environment that helps to create the commodity. Thus, they interrogate the Chilean case, which is recognised worldwide for its salmon exports, to evaluate if the environmental effects of such industry can be reduced to zero. Despite some ongoing initiatives being implemented in the salmon resource peripheries, they unveil several barriers currently impeding the sustainable implementation of the circular economy model. Some of these are related to institutional rigidities, the features of the industry, and the cultural understanding of waste and recycling.

Finally, Miguel Atienza's chapter focuses on how labour has been geographically fragmented in extractive global production networks and will be increasingly fragmented in the near future due to the "third wave of globalization". This process comprises stark shifts in the distribution and location of the working force due to the increasing automation, long-distance commuting (fly-in fly-out and shift systems) and remote work. Hence, this chapter proposes expanding the agenda on resource peripheries to re-examine the assumption that labour is spatially fixed next to the extraction/exploitation sites, by better incorporating issues related to value capture, territorial and network embeddedness and the power asymmetries and exertion shaping the developmental outcomes that resource peripheries are currently facing. It is by incorporating these recent shifts, Atienza claims, that we could better re-distribute the uneven capture of value by lead firms and non-local workers to the losing resource peripheries

Acknowledgements Felipe Irarrázaval is grateful to the Centro de Estudios Sobre Conflicto y Cohesión Social (COES) (ANID/FONDAP/15130009)

References

Allen J (2016) Topologies of power: beyond territory and networks. Routledge, New York

Alonso-Fradejas A Jr (2015) Food sovereignty: convergence and contradictions, conditions and challenges. Third World Quart 36(3):431–448. https://doi.org/10.1080/01436597.2015.1023567

Arboleda M (2016) Spaces of extraction, metropolitan explosions: planetary urbanization and the commodity boom in Latin America. Int J Urban Reg Res 40(1):96–112. https://doi.org/10.1111/1468-2427.12290

Arboleda M (2020) Planetary mine: territories of extraction under late capitalism. Verso, New York

Arias-Loyola M (2021) Evade neoliberalism's turnstiles! lessons from the Chilean Estallido social. Environ Plan A

Arias M (2014) Large mining enterprises and regional development in Chile: between the enclave and cluster. J Econ Geogr 14(1):73–95. https://doi.org/10.1093/jeg/lbt007

Argent N (2013) Reinterpreting core and periphery in Australia's mineral and energy resources boom: an Innisian perspective on the Pilbara. Aust Geogr 44(3):323–340. https://doi.org/10.1080/00049182.2013.817033

Argent N (2017) Rural geography I: resource peripheries and the creation of new global commodity chains. Prog Hum Geogr 41(6):803–812. https://doi.org/10.1177/0309132516660656

Atienza M (2020) Gateways or backdoors to development? filtering mechanisms and territorial embeddedness in the Chilean copper GPN's urban system. Growth Chang 52(1):88–110. https://doi.org/10.1111/grow.12447

Atienza M, Fleming D, Aroca P (2020) Territorial development and mining. Insights and challenges from the Chilean case. Resour Policy 101–812. https://doi.org/10.1016/j.resourpol.2020.101812

Baglioni E, Campling L (2017) Natural resource industries as global value chains: frontiers, fetishism, labour and the state. Environ Plan A: Econ Space 49(11):2437–56. https://doi.org/10.1177/0308518X17728517

Bair J (2011) Commodity chains and the uneven geographies of global capitalism: a disarticulations perspective: Environ Plan A. https://doi.org/10.1068/a43505

Bair J (2009) Global commodity chains: genealogy and review in: frontiers of commodity chain research. In: Bair J (ed) Frontiers of commodity chain research. Stanford University Press

Barratt T (2019) Temporality and the evolution of GPNs: remaking BHP's Pilbara iron ore network. Reg Stud 1–10. https://doi.org/10.1080/00343404.2019.1590542

Barrientos S (2014) Gendered global production networks: analysis of cocoa–chocolate sourcing. Reg Stud 48(5):791–803

Bebbington A (2018) Governing extractive industries: politics, histories, ideas. Oxford University Press, Oxford

Bebbington A (2008) Mining and social movements: struggles over livelihood and rural territorial development in the andes. World Dev 36(12):2888–2905. https://doi.org/10.1016/j.worlddev.2007.11.016

Barton J (2008) Transformations in resource peripheries: an analysis of the Chilean experience. Area 40(1):24–33. https://doi.org/10.1111/j.1475-4762.2008.00792.x

Brenner N (2004) New state spaces: urban governance and the rescaling of statehood. Oxford University Press USA—OSO. http://ebookcentral.proquest.com/lib/manchester/detail.action?docID=422528

Breul M (2018) An intermediate step to resource peripheries: the strategic coupling of gateway cities in the upstream oil and gas GPN. Geoforum 92:9–17. https://doi.org/10.1016/j.geoforum.2018.03.022

Breul M (2021) One thing leads to another, but where?—gateway cities and the geography of production linkages. Growth Chang 52(1):29–47. https://doi.org/10.1111/grow.12347

Bridge G (2008) Global production networks and the extractive sector: governing resource-based development. J Econ Geogr 8(3):389–419. https://doi.org/10.1093/jeg/lbn009

Bridge G (2020) New energy spaces: towards a geographical political economy of energy transition. Environ Plan A: Econ Space 52(6):1037–1050. https://doi.org/10.1177/0308518X20939570

Brosio G, and Jiménez JP (2012) Decentralisation and reform in Latin America: improving intergovernmental relations. Edward Elgar Publishing

Carson DB, Carson DA, Porter R, Ahlin CY and Sköld P (2016) Decline, adaptation or transformation: new perspectives on demographic change in resource peripheries in Australia and Sweden. Comp Popul Stud 41(3–4), Article 3–4. https://comparativepopulationstudies.de/index.php/CPoS/article/view/245

Ciccantell P (2009) Rethinking global commodity chains: integrating extraction, transport, and manufacturing. Int J Comp Sociol 50(3–4):361–384. https://doi.org/10.1177/0020715209105146

Coe N (2004) 'Globalizing' regional development: a global production networks perspective. Trans Inst Br Geogr 29(4):468–484. https://doi.org/10.1111/j.0020-2754.2004.00142.x

Coe N, Yeung HWC (2015) Global production networks: theorizing economic development in an interconnected world. Oxford University Press

Le Conde M (2017) Why do some communities resist mining projects while others do not? Extr Ind Soc 4(3):681–697. https://doi.org/10.1016/j.exis.2017.04.009

Cust J (2015) The local economic impacts of natural resource extraction. Ann Rev Resource Econ 7(1):251–268. https://doi.org/10.1146/annurev-resource-100814-125106

Delamaza G, Maillet A, Neira CM (2017) Socio-territorial conflicts in Chile: configuration and politicization (2005–2014). Eur Rev Latin Am Carib Stud 23–46. https://doi.org/10.18352/erlacs.10173

Ellem B (2015) Resource peripheries and neoliberalism: the Pilbara and the remaking of industrial relations in Australia. Aust Geogr 46(3):323–337. https://doi.org/10.1080/00049182.2015.1048587

Fry M (2018) Petro-geographies and hydrocarbon realities in Latin America. J Lat Am Geogr 17(3):10–14

Gago V (2017) A critique of the extractive operations of capital: toward an expanded concept of extractivism. Rethink Marx 29(4):574–591. https://doi.org/10.1080/08935696.2017.1417087

Gereffi G (2018) The emergence of global value chains: ideas, institutions, and research communities. In: Global value chains and development: redefining the contours of 21st century capitalism, pp 1–40. Cambridge University Press. https://doi.org/10.1017/9781108559423.002

Gibson-Graham JK (1996) The end of capitalism (as we knew it): a feminist critique of political economy. Blackwell, Oxford

Gudynas, E. (2010). The new extractivism of the 21st century: ten urgent theses about extractivism in relation to current South American progressivism. Americas Program Report, 21, 1–14. In: Haarstad H (2014) Cross-scalar dynamics of the resource curse: constraints on local participation in the Bolivian gas sector. J Dev Stud 50(7):977–990. https://doi.org/10.1080/00220388.2014.909026

Haarstad H (2014) Cross-scalar dynamics of the resource curse: constraints on local participation in the Bolivian gas sector. J Dev Stud 50(7):977–990. https://doi.org/10.1080/00220388.2014.909026

Hayter R (2003) The war in the woods: post-fordist restructuring, globalization, and the contested remapping of British Columbia's forest economy. Ann Assoc Am Geogr 93(3):706–729. https://doi.org/10.1111/1467-8306.9303010

Hayter R (2012) Neoliberalization and its geographic limits: comparative reflections from forest peripheries in the global north. Econ Geogr 88(2):197–221. https://doi.org/10.1111/j.1944-8287.2011.01143.x

Hayter R (2003) Relocating resource peripheries to the core of economic geography's theorizing: rationale and agenda. Area 35(1):15–23. https://doi.org/10.1111/1475-4762.00106

Hayter R (2000) Single industry resource towns. In: Sheppard E, Barnes TJ A companion to economic geography, pp 290–307. John Wiley & Sons, Ltd. https://doi.org/10.1002/9781405166430.ch18

Irarrázaval F (2021) Natural gas production networks: resource making and interfirm dynamics in Peru and Bolivia. Ann Assoc Am Geogr 111(2):540–558

Irarrazaval F (2020) Contesting uneven development: the political geography of natural gas rents in Peru and Bolivia. Polit Geogr 79. https://doi.org/10.1016/j.polgeo.2020.102161

Kanai JM (2016) The pervasiveness of neoliberal territorial design: cross-border infrastructure planning in South America since the introduction of IIRSA. Geoforum 69:160–170. https://doi.org/10.1016/j.geoforum.2015.10.002

Kirshner J (2015) Mining and extractive urbanism: postdevelopment in a Mozambican boomtown. Geoforum 61:67–78. https://doi.org/10.1016/j.geoforum.2015.02.019

Markusen A (1996) Sticky places in slippery space: a typology of industrial districts. Econ Geogr 72(3):293–313. JSTOR. https://doi.org/10.2307/144402

Massey D (1984) Spatial divisions of labour: social structures and the geography of production. Macmillan, London

McElroy C (2018) Reconceptualizing resource peripheries. The New Oxford Handbook of Economic Geography. https://doi.org/10.1093/oxfordhb/9780198755609.013.32

MacLeod G (2001) New regionalism reconsidered: globalization and the remaking of political economic space. Int J Urban Reg Res 25(4):804–829. https://doi.org/10.1111/1468-2427.00345

McGrath S (2018) Dis/articulations and the interrogation of development in GPN research. Prog Hum Geogr 42(4):509–528. https://doi.org/10.1177/0309132517700981

Morris M (2012) One thing leads to another—commodities, linkages and industrial development. Resour Policy 37(4):408–416. https://doi.org/10.1016/j.resourpol.2012.06.008

Nelson R (2004) The Peripheries of British Columbia: patterns of migration and economie structure, 1976–2002. Can J Reg Sci 27(3):353–394

Perez C (2015) The new context for industrializing around natural resources: an opportunity for Latin America (and other resource rich countries)? (No 62; the other Canon Foundation and Tallinn University of Technology Working Papers in Technology Governance and Economic Dynamics). TUT Ragnar Nurkse Department of Innovation and Governance. https://ideas.repec.org/p/tth/wpaper/62.html

Perreault T (2018) Energy, extractivism and hydrocarbon geographies in contemporary Latin America. J Lat Am Geogr 17(3):235–252

Phelps NA (2015) Encore for the enclave: the changing nature of the industry enclave with illustrations from the mining industry in Chile. Econ Geogr 91(2):119–146. https://doi.org/10.1111/ecge.12086

Phelps NA (2018) An invitation to the dark side of economic geography. Environ Plan A: Econ Space 50(1):236–244. https://doi.org/10.1177/0308518X17739007

Rasmussen MB (2018) Reconfiguring frontier spaces: the territorialization of resource control. World Dev 101:388–399. https://doi.org/10.1016/j.worlddev.2017.01.018

Rehner J (2020) Boom city! regional resource peripheries and urban economic development in Chile. Area Dev Policy 5(3):305–323

Rich JAJ (2019) Introduction the politics of participation in Latin America: new actors and institutions. Latin Am Politics Soc 61(2):120. https://doi.org/10.1017/lap.2018.74

Riofrancos TN (2017) Scaling democracy: participation and resource extraction in Latin America. Perspect Polit 15(3):678–696. https://doi.org/10.1017/S1537592717000901

Scholvin S, Breul M, Diez JR (2019) Revisiting gateway cities: connecting hubs in global networks to their hinterlands. Urban Geogr 0(0):1–19. https://doi.org/10.1080/02723638.2019.1585137

Scholvin S (2020) World cities and peripheral development: the interplay of gateways and subordinate places in Argentina and Ghana's upstream oil and gas sector. Growth Change, Grow 1:23–86. https://doi.org/10.1111/grow.12386

Springer S (2012) Neoliberalising violence: of the exceptional and the exemplary in coalescing moments. Area 44(2):136–143

Steen M (2011) Upgrading the 'Petropolis' of the North? resource peripheries, global production networks, and local access to the Snøhvit natural gas complex. Norsk Geografisk Tidsskrift Norw J Geogr 65(4):212–225. https://doi.org/10.1080/00291951.2011.623307

Stephansen HC (2013) Connecting the peripheries: networks, place and scale in the world social forum process. J Postcolonial Writ 49(5):506–518

Suarez-Cao J (2017) Presentación: El auge de los estudios sobre la política subnacional latinoamericana. Colomb Int 90:15–34. https://doi.org/10.7440/colombiaint90.2017.01

Svampa M (2019) Las fronteras del neoextractivismo en América Latina: conflictos socioambientales, giro ecoterritorial y nuevas dependencias. Bielefeld Univeristy Press, Germany

Swyngedouw E (2004) Globalisation or 'glocalisation'? networks, territories and rescaling. Camb Rev Int Aff 17(1):25–48. https://doi.org/10.1080/0955757042000203632

Temper L (2020) Movements shaping climate futures: a systematic mapping of protests against fossil fuel and low-carbon energy projects. Environ Res Lett 15(12). https://doi.org/10.1088/1748-9326/abc197

Tsing AL (2011) Friction: an ethnography of global connection. Princeton University Press

De Vesco P (2020) Natural resources and conflict: a meta-analysis of the empirical literature. Ecol Econ 172. https://doi.org/10.1016/j.ecolecon.2020.106633

Werner M (2016) Global production networks and uneven development: exploring geographies of devaluation, disinvestment, and exclusion. Geogr Compass 10(11):457–469. https://doi.org/10.1111/gec3.12295

Werner M (2019) Geographies of production I: global production and uneven development. Prog Hum Geogr 45(5):948–958. https://doi.org/10.1177/0309132518760095

Wilson J (2011) Colonising space: the new economic geography in theory and practice. New Polit Econ 16(3):373–397. https://doi.org/10.1080/13563467.2010.504299
World Bank and WTO (2019) Global value chain development report 2019: technological innovation, supply chain trade, and workers in a globalized world (No 136044; pp 1–196). The World Bank. http://documents.worldbank.org/curated/en/384161555079173489/Global-Value-Chain-Development-Report-2019-Technological-Innovation-Supply-Chain-Trade-and-Workers-in-a-Globalized-World

Felipe Irarrázaval is postdoctoral researcher and adjunct professor at the Instituto de Estudios Urbanos y Territoriales at the Pontificia Universidad Católica de Chile and at the Center for the Study of Conflicts and Social Cohesion (COES). He received a Ph.D. in Human Geography from The University of Manchester. His research examines resource governance, particularly extractive industries in Latin America, through the lens of global production networks, political geography and urban studies. He has undertaken research in Peru, Bolivia and Chile, and has published his research in journals like Economic Geography, Political Geography, Annals of AAG, Capitalism Nature and Socialism, and EURE.

Martín Arias-Loyola has received a Ph.D. in Economic Geography and Planning Studies from the University College London (UK) and M.Phil. in Regional Sciences from the Universidad Católica del Norte (Chile). He is Assistant Professor of Economic Geography at the Departamento de Economía and researcher at the IDEAR of the Universidad Católica del Norte in Antofagasta (Chile). He is also a Visiting Academic/Postdoctoral Researcher at the Faculty of Architecture, Building and Planning at the University of Melbourne (Australia). His research focuses on power asymmetries; the dark side of economic, political, and urban geography; extractivism; neoliberal extractive contexts; cooperativism; urban politics; mutual aid; informal settlements; and the role of academia in social change and the right to the city.

Part I
Networks

Chapter 2
Commodity Chains and Extractive Peripheries: Coal and Development

Paul S. Ciccantell

Abstract This chapter will utilize the raw materialist lengthened global commodity chains model to examine how the coal global commodity chain has shaped extractive peripheries over the last three centuries. Coal is the quintessential "old economy" raw material that powered the Industrial Revolution. Coal was a key ingredient in economic ascent over the past three centuries in Great Britain, the U.S., and Japan and remains essential to economic ascent in the twenty-first century for China and India, despite its contribution to climate change and efforts in many countries to promote a transition toward more sustainable energy systems. After presenting the theoretical model guiding this analysis, the chapter will outline how the role of coal commodity chains has changed over time as they have grown in scale from small localized chains to truly global chains, highlighting the roles that the global coal commodity chain plays in the world economy today and the drivers of change in the coal industry. How do raw materialist lengthened global commodity chains based on the extraction, processing, and consumption of coal shape the extractive peripheries that supply coal?

Keywords Coal · Global commodity chains · Economic ascent · Extractive peripheries · Resource frontiers · Raw materialism

2.1 Introduction

This chapter will utilize the raw materialist lengthened global commodity chains model (Ciccantell and Smith 2009; Ciccantell and Gellert 2018; Gellert and Ciccantell 2020) to examine how the coal global commodity chain has shaped extractive peripheries and their potential for development over the last three centuries. Coal is the quintessential "old economy" raw material that powered the Industrial Revolution. Coal was a key ingredient in economic ascent over the past three centuries in Great Britain, the U.S., and Japan (Bunker and Ciccantell 2005, 2007) and remains

P. S. Ciccantell (✉)
Department of Sociology, Western Michigan University, Kalamazoo, USA
e-mail: paul.ciccantell@wmich.edu

F. Irarrazaval and M. Arias-Loyola (eds.), *Resource Peripheries in the Global Economy*, Economic Geography, https://doi.org/10.1007/978-3-030-84606-0_2

essential to economic ascent in the twenty-first century for China and India (Ciccantell 2009; Ciccantell and Gellert 2018; Gellert and Ciccantell 2020), despite its contribution to climate change and efforts in many countries to promote a transition toward more sustainable energy systems.

To contribute to the goals of this volume, this chapter will examine the evolution of the coal commodity chain over time to uncover how coal resource peripheries are produced through dynamic and integrated socio-spatial relationships. The chapter aims to analyze the ways variegated socio-spatial relations underpinning the global economy are still producing coal resource peripheries that are essential for core and ascendant economies' functioning. The key element is raw materialist lengthened global commodity chains that link core and ascendant economies to resource peripheries, a socio-spatial relationship that creates and reproduces global inequalities and shapes the socioeconomic trajectories of those locations that become resource peripheries. This approach traces the steps of processing and the transport and communication technologies that link the multiple nodes of the commodity chain from its raw materials/agricultural sources through industrial processing to consumption and eventually waste disposal. This materially and spatially grounded approach allows analysis of the economic, social, and environmental dimensions of these chains at each node.

Recent work in economic geography using the global value chain (GVC) framework seeks to integrate the upstream end of commodity chains and the material characteristics of resources, their extraction, and processing and consumption into their analysis as essential factors that shape these chains, while recapturing the critical stance of earlier approaches that recognized that incorporation into these chains may lead not only to development but also to underdevelopment, beginning with the work of Bridge (2008). Havice and Campling (2017) and Campling and Havice (2019) used the tuna global value chain as a case study of how to rework this approach to take seriously "nature" and materiality. Their papers and others in this tradition focused on the salmon industry (Irarrázaval and Bustos-Gallardo 2019), natural gas (Bridge and Bradshaw 2017), and natural resources more generally (Baglioni and Campling 2017) all bring resources into a central place in this analysis, arguing that the understanding of chain governance needs to expand beyond efforts to create typologies of chains to include an explicit emphasis on environmental conditions of production and environmental problems (Havice and Campling 2017). The careful analysis of the materiality of these commodities and of the firm strategies and governance issues across the value chains and contestation between actors provides the sort of rich, grounded understanding of a chain that begins with a "natural resource" that is fundamentally shaped by the relationship between nature and society, materiality, and space as location and territory (Havice and Campling 2017; Campling and Havice 2019; Baglioni and Campling 2017; Bridge and Bradshaw 2017; Irarrázaval and Bustos-Gallardo 2019) that the raw materialist lengthened GCC approach seeks to build. This reformulated GVC approach further shares a methodological emphasis on developing deep familiarity with the chains under study via combining interviews, observations, and quantitative data using both data and methodological triangulation (Yin 2017). This approach also seeks to extend the analytic focus to a much longer term (more historical) than is typical for GVC analysis (see also Ciccantell Forthcoming for a related effort to bring together sociological and geographic perspectives on natural resources).

Following the theme of this volume, this chapter responds to Hayter et al.'s (2003) call for theorizing about resource peripheries. Hayter et al. (2003) emphasized the diversity of resource peripheries and the role of a few key processes in shaping them, focusing on four sets of institutional values or dimensions, including industrialism, environmentalism, aboriginalism, and imperialism (Hayter et al. 2003: 17). Hayter et al. (2003) also argued that resource peripheries are contested places, with this contestation integral to imperialism and globalization, emphasizing that "these fears were that resource exploitation directed by outside powers was not in the best local interest…in geopolitical terms at least, resource peripheries are some of the most contested parts of the world" (Hayter et al. 2003: 19), a concern my theoretical approach shares. However, no comprehensive definition of resource peripheries is offered by Hayter et al. (2003) to begin developing the theoretical understanding they call for in the article. My approach in this chapter shares the empirical focus of Havice and Campling (2017; Campling and Havice 2019), Bridge (2008) and colleagues interested in extractivism in Latin America and other areas and Hayter et al. (2003) in economic geography. However, my theoretical model is rooted in world-systems theory as reformulated in Bunker and Ciccantell (2005, 2007) and Ciccantell and Smith (2009).

In a forthcoming publication (Ciccantell and Gellert 2021), I define extractive peripheries as locations in the capitalist world-economy whose primary economic role in the world-system is to supply some type of raw material via commodity chains to other locations that process and/or consume these raw materials. An extractive periphery is an explicitly geographical and time-bound concept. It can be located within a peripheral nation (e.g., sugar plantations in Haiti in the 1600s and 1700s), a semi-peripheral nation (e.g., rainforest converted to farm and ranch land in the Amazon in Brazil since the 1960s), or even a core nation (e.g., coal mining areas of Appalachia in the U.S. since the late 1800s). Before a particular location becomes an extractive periphery, it must first be transformed into a resource frontier. A resource frontier is the first stage of transforming an "external arena" (Wallerstein 1974; Chase-Dunn and Hall 1997) into an extractive periphery that is "new" and in the process of being incorporated into the world-system on the basis of a particular raw material, with existing populations losing access to and control over land and other resources. The area's ecosystems and landscapes in a resource frontier are in the process of being redefined and reconstructed over time to serve the interests of wealthier and more powerful groups in distant locations. In many cases, a location becomes a resource frontier as outside groups invade in search of valuable resources. In other cases, a location can be restructured from an extractive periphery for one purpose into a (new) resource frontier for another purpose (e.g., the shift of much of northern and central Appalachia from a coal extractive periphery to a resource frontier for natural gas via hydraulic fracturing in the past decade) (Ciccantell and Gellert 2021).

After presenting the theoretical model guiding this analysis, the chapter will outline how the role of coal commodity chains has changed over time as they have grown in scale from small localized chains to truly global chains, highlighting the roles that the global coal commodity chain plays in the world economy today and the drivers of change in the coal industry. The chapter then focuses on the consequences for extractive peripheries of incorporation into this global commodity chain driven by core-periphery relations and the power of states and firms in core nation-states and in rising economies that seek to reshape the world economy to support their economic ascent. How do raw materialist lengthened global commodity chains (Ciccantell and Smith 2009) based on the extraction, processing, and consumption of coal shape the extractive peripheries that supply coal?

2.2 Natural Resources, Development, and Raw Materialist Lengthened Global Commodity Chains

Natural resources were historically seen as natural capital that could be extracted using technology, capital, and labor to produce economic growth and development by unlocking and putting into circulation this natural produced capital (Hirschmann 1958, 1977; Perroux 1955; Perloff and Wingo 1961; Watkins 1963; Bunker 1989). However, this historical view has been criticized from a variety of perspectives, both in economic terms and more broadly in terms of social and environmental sustainability (Bunker 1989; Bunker and Ciccantell 2005; Bebbington et al. 2018; Arboleda 2020).

One frequent criticism of the impacts of natural resource dependence on national and local economic development is the concept of the "resource curse", the view that exploitation of natural resource endowments leads to negative economic and social outcomes (see Ciccantell and Patten 2016). Overall, despite the popular reception that the concept of the resource curse has received, actual empirical evidence is at best quite mixed (Ciccantell and Patten 2016). A different line of criticism focuses on the local impacts of extraction. Raw material extraction is generally regarded as an uncertain and unstable basis for economic development due to boom and bust cycles (Bunker 1985; Freudenburg 1992; Freudenburg and Wilson 2002; Wilson 2001, 2004; Krannich et al. 2014), with one analyst referring to the situation as "riding the resource roller coaster" (Wilson 2001, 2004). The extractive enclaves where natural resources are obtained from nature are often company towns owned by foreign or national firms with headquarters and other operations hundreds or thousands of miles away. These enclaves capture few of the benefits of extraction, create few linkages to other forms of local economic activity and are often abandoned when the "resource roller coaster" crashes due to resource exhaustion, low prices in global markets, firm strategies that disarticulate the enclave from its commodity chain due to changes in government taxation or other policies, or simply because another extractive location offers some benefit to the extractive firm (extractive enclaves and company towns are a longstanding phenomenon of interest in development studies, geography and sociology; see, e.g., among many others, Lucas 1971 on Canada, the papers in Dinius and Vergara 2011 that examine cases across the Americas and the extensive bibliography provided, Green 2010 on the U.S, Gaventa 1980 on U.S. coal towns, Nash's 1979 classic on Bolivia, and, more recently, Phelps, Atienza and Arias Loyola 2018 on uneven and negative aspects of economic geography, and Irarrázaval and Bustos-Gallardo 2019 on salmon).

Coal is a classic case of a staple, a natural resource that is extracted from nature and exported to an industrial nation and that shapes the economic, social and environmental history of the extractive region (Innis 1956). Mining coal is tied to a particular and often socially remote location determined by nature. Transporting coal to the locations of human use is very expensive, typically requiring railroad or water transport to make it affordable. The lack of backhaul cargo on this coal transport system increases the costs of transport and the risks of investing in the required transport system. The lack of labor in socially remote naturally determined locations of coal is a key obstacle and expense for firms and states that seek to utilize the coal to earn a profit and promote economic development. Lowering labor needs and costs via technological innovations of increasing scale, reorganization of work processes, and lowering skill levels of workers to reduce labor costs are all priorities for coal mining firms. Coal mining inevitably leads to damage of the natural environment during extraction, initial processing, transport, and consumption.

Despite these difficult challenges, staples in general and coal in particular have long been seen as key ingredients of industrialization and economic development. Innis' (1956) empirically grounded analysis of Canada's economic history, highlighting both the positive and negative impacts of staples production in shaping this history, was transformed into a theory and model of economic development that used the roles of coal, iron ore, timber, and other raw materials in the industrialization of Canada, Great Britain, and the U.S. as the empirical evidence that raw materials could provide comparative advantages for building industries such as steel, railroads, and shipbuilding. Raw materials provided the basis for constructing growth poles (Perroux 1955) and linkages from mines to smelters to modern industry via railroads, ships, and trucks (Hirschmann 1958); Watkins (1963) explicitly called his version a staple theory of economic growth. Modernization theory (Rostow 1960) and resource-based development theory (Auty and Mikesell 1999) explicitly or implicitly rested on this same foundation of raw materials as a basis for economic development. From the dependency literature of the 1960s (Frank 1967; Cardoso and Faletto 1969) to world-systems analysis (Wallerstein 1974; Bunker and Ciccantell 2005, 2007) to critical economic geography in recent years (Campling and Havice 2019; Havice and Campling 2017; Arias Loyola et al. 2014), the positive assumptions about natural resource extraction and development have long been critically challenged.

How can this contradictory body of evidence be understood? One approach is to situate particular cases of dependence on raw materials extraction within the broader processes of long-term change in the world economy. The raw materialist lengthened global commodity chains theoretical and methodological model brings together the global commodity chains model (Hopkins and Wallerstein 1986; Gereffi and Korzeniewicz 1994; Bair 2005, 2009) and new historical materialism (Bunker and Ciccantell 2005, 2007), or to put it more bluntly, raw materialism. The raw materialist model begins from a focus on the material process of economic ascent in the capitalist world-economy, defined as "the development of increasing economic, political and military power relative to competing states and the existing hegemon" (Ciccantell 2020a,b: 6) on the part of a state and its economy that are growing rapidly toward core or even hegemonic status. This materially grounded analysis of the most dramatic cases of economic ascent over the last four centuries (Great Britain, the U.S., Japan, and China) revealed a consistent pattern: their rapid development of larger, more complex economies and powerful states and firms capable of restructuring many parts of the world economy and polity in support of this growing power rested on a foundation of a few fundamental raw materials based industries, most critically the iron and steel industries and the transport industries to move essential raw materials at lower cost and on a larger scale than the existing hegemon, other core economies, and other competing ascendant economies.

The key problem for rapidly growing economies over the past five centuries has been obtaining raw materials in large and increasing volumes to supply their continued economic development in the context of economic and geopolitical cooperation and conflict with the existing hegemon and other rising economies. Economies of scale offer opportunities to reduce costs and create competitive advantages relative to the existing hegemon and other rising economies, but raw materials depletion and increasing distance create diseconomies of space (increasing costs due to the need to bring raw materials from ever more distant extractive peripheries to the consuming regions) that make finding economic, technological, and sociopolitical fixes via increasing economies of scale difficult to achieve, maintain, and eventually reconstruct on an even larger scale (Bunker and Ciccantell 2005, 2007).

Successfully resolving this contradiction relies on the creation of generative sectors. Generative sectors create backward and forward linkages; create patterns of relations between firms, sectors and states; stimulate a range of technical skills and learning and social institutions to fund and promote them; and stimulate creation of a financial system to meet complex and costly capital needs across borders. In short, generative sectors drive economic ascent. Building these generative sectors is a highly contentious and tenuous process that must be maintained in dynamic tension; it is far more common for efforts in rising economies to create and maintain these sectors to fail than to succeed. The U.S., Germany, and France all built one of the largest steel industries in the late 19th and early twentieth centuries as they sought to challenge British hegemony, but only the U.S. was able to maintain in dynamic tension the material, financial, and political organization that sustained economic ascent to a hegemonic position (Bunker and Ciccantell 2005, 2007).

These processes of economic ascent and economic and geopolitical competition with existing hegemons have driven long term change in the capitalist world-economy over the past five centuries (Bunker and Ciccantell 2005). The most dramatic and rapid processes of economic ascent restructure national economies and the world economy in support of national economic ascent. The competitive advantages created by organizational and technological innovations in generative sectors and by subsidies from peripheries lead to global trade dominance. Economic and political competition from the existing hegemon and other ascending economies shapes and constrains long term success, making economic ascent and challenges to existing hegemons extremely difficult. The most successful cases of ascent restructure and progressively globalize the world economy, incorporating and reshaping economies, ecosystems, and space. The historical sequence of rapidly ascending economies from Holland to Great Britain to the U.S. to Japan led to dramatic increases in the scale of production and trade, building generative sectors in iron and steel, petroleum, railroads, ocean shipping, and other raw materials and transport industries that drove their economic ascent and impoverished their raw materials peripheries (Bunker and Ciccantell 2005, 2007).

These processes of long-term structural change, the contradiction between economies of scale and diseconomies of space, and the challenges facing the creation and reproduction of generative sectors raise the bar for future ascendant economies. One central element of the challenges faced by ascendant economies is the recognition that technological and organizational innovations in ascendant economies that resolve this tension simultaneously benefit the rising economy and impoverish its raw material peripheries, increasing global inequality. Underdevelopment of the periphery (Frank 1967; Cardoso and Faletto 1969; Bunker 1985) is an inherent element of the development of ascendant economies and of existing hegemons. As the dependency and world-system theorists (Frank 1967; Cardoso and Faletto 1969; Wallerstein 1974; Rodney 1982; Bunker 1985, among many others) have emphasized from the 1960s onward, the economic, military, and political power inequalities between powerful core nation-states and their peripheries in Latin America, Africa, and Asia have driven the underdevelopment of the periphery and the multitude of negative consequences of participation in commodity chains and broader patterns of economic, colonial, political, and other connections for the people of the periphery.

The newer ascendants' rapid growth, however, means that their raw materials demands are increasing dramatically and necessitating a similar rise in supply if these growth rates are to be sustained. The combination of the existing social and material infrastructures in the raw materials peripheries established by earlier ascendants, rapid demand growth in the newly ascendant economy, and the willingness of the newer ascendant economy to pay higher prices for raw materials in order to sustain their domestic growth creates an opportunity that states and firms in the raw materials periphery find very attractive. Higher prices for rapidly increasing volumes of exports (in contrast to slower demand growth in the mature economies of the earlier ascendant) motivate firms and domestic elites in the periphery and even from existing core powers with fewer opportunities for profitable investments to invest in production for export to the new ascendant. States in raw materials exporting regions typically support this investment with subsidies for transport and extraction, both in an effort to promote economic development and in hopes of gaining better returns and more political freedom from the power of the existing hegemon. This is particularly apparent in postcolonial situations in which newly independent states seek to break free from neocolonial ties and in situations of resource nationalism in which states seek greater control over and benefits from raw materials exports. Firms, elites, and states in raw materials peripheries come to see the new ascendant as a potential ally in their attempts to promote political independence and economic development (Ciccantell 2009).

From the perspective of the new ascendant, building these relationships with existing raw material peripheries is much less expensive and difficult than creating their own new peripheries. One of the most important benefits is that most of the cost and the risk of expanding extraction and transport is borne by firms and states in the extractive periphery and often by firms from the earlier ascendant. At the same time, these investments in mines and transport systems also often create opportunities for exports of industrial products from the ascendant economy to the periphery to support the development of these extractive industries and for consumption by the owners of and workers in these industries. "Stealing" these peripheries from earlier ascendants thus further enhances the rapid growth of the new ascendant by reducing costs and risks while simultaneously creating significant new opportunities for profit from trade and investment (Ciccantell 2009).

Creating these commodity chains and their resultant inequalities depends critically on the creation of transport networks that can move raw materials from their naturally determined locations in extractive peripheries to industries and consumers in newly ascendant and core economies. In a very material sense, transport systems are the sinews of hegemonic rivalry and stealing extractive peripheries. Railroads, ports, pipelines, and oceangoing ships have moved progressively larger amounts of raw materials from increasingly distant extractive peripheries over the past five centuries in the capitalist world-economy (Bunker and Ciccantell 2005, 2007).

Equally important, this approach provides a lens to examine spatially-based disarticulations (the marginalization or outright elimination of particular nodes from a GCC, such as via closure of a factory) (Bair and Werner 2011; Werner 2015, 2018) and contestations over extraction, processing, transport, consumption, and waste disposal across these chains. This grounded analysis can examine development trajectories and the sociopolitical conflicts over the division of costs and benefits in particular nodes and across these commodity chains. This approach highlights the role of contestation and resistance to the construction and reproduction of a particular commodity chain in particular places, as, for example, labor movements and social movement organizations seek to achieve their goals despite resistance from firms and states that oppose these goals, as we have analyzed in a variety of commodity chains, including coal, oil and gas, aluminum, steel, and textile manufacturing (Ciccantell and Smith 2009; Sowers et al. 2014, 2017, 2018; Ciccantell 2020a, b and Forthcoming).

This model thus emphasizes long term historical change in the world-system as a whole and in particular places and times, and it allows world-systemic comparative analysis that makes nested and over time comparisons across commodity chains. The grounding in material process also focuses attention on local, regional, and global environmental impacts of these lengthened global commodity chains, a sometimes neglected dimension of world-systems analysis.

Overall, this raw materialist lengthened global commodity chains approach provides an integrated theoretical and methodological approach to examine the impacts of particular commodity chains both in specific times and places, including particular rural extractive peripheries, and as the constitutive elements of long term change in the capitalist world-economy.

2.3 The Evolution of Coal Commodity Chains Over the Long Term

2.3.1 The Materiality of Coal Shapes Coal Commodity Chains

Coal became the most important source of industrial power by the beginning of the 1800s (Malm 2016), a role it maintained through the mid-twentieth century and it remains surprisingly central in the twenty-first century (Ciccantell and Gellert 2018; Gellert and Ciccantell 2020). The materiality of coal shaped this long-term essential role of coal commodity chains in the world economy in a number of ways. From the raw materialist lengthened GCC framework, the material and social roles of coal over the last three centuries became fundamental elements of the evolution of the capitalist world-economy and long-term social change. The coal commodity chains that fueled these long-term processes shaped internal and external extractive peripheries that supplied coal to hegemons and ascending economies.

Coal did not disappear after Great Britain's economic ascent or after the first decades of U.S. ascent; it played a critical role in Japan's post-World War II ascent and in China's ascent since the 1980s (Bunker and Ciccantell 2005, 2007). Despite its declining importance in the U.S. and much of Europe today, coal remains a critical component of economic growth in the fastest growing ascendant economies, China, and India. In short, earlier materials and generative sectors do not disappear over the *longue durée* as systemic cycles of accumulation move forward (Ciccantell and Gellert 2018; Gellert and Ciccantell 2020).

This long term and continued critical role of coal has been fundamentally shaped by the material characteristics of coal. Most importantly, coal burns and produces heat and thereby power in various forms. This heat and the release of carbon from burning can smelt metal ores into forms more useful to humans (Harris 1988; Isard 1948; Pomeranz 2000; Wu 2015; McGraw-Hill 1992: 50). All types of coal can be burned to generate electricity (coal used for this is termed thermal or steam coal). Different coals are differentially efficient in generating heat and electricity, but this does not affect the quality of the final product. It is relatively easy to switch coal suppliers within the same major type of coal without any significant alterations in plant equipment or operations. This relatively easy substitutability of thermal coals has fostered competition between firms and between coal extracting countries and facilitated the globalization of the coal industry (Ciccantell and Gellert 2018).

In contrast, only a very limited range of bituminous coals with very particular qualities can be used to smelt iron ore and produce steel because coal quality directly affects the quality of the metal produced (McGraw-Hill 1992: 49–60; Grainger and

Gibson 1981: 11–16, 130–139). As a result, the substitutability of metallurgical coal from different deposits is significantly more difficult than it is for thermal coal. In order to run steel mills at full capacity and without expensive shutdowns, stability of supply is essential. Firms and states in core and ascendant economies seek to either own or create long term contracts with their coal sources in extractive peripheries, creating coal commodity chains that are tightly integrated over long periods of time. Internal coal peripheries are preferred because of geopolitical security, but growing demand and resource depletion often force firms and states to extend commodity chains to external coal peripheries. Great Britain and the U.S. enjoyed secure internal coal peripheries, but Japan and China had to build external coal commodity chains linked to external extractive peripheries to support their economic ascent based on steel.

In contrast to the vast majority of minerals, geologic and environmental processes over millions of years produced huge volumes of various types of coals in many locations spread around the globe; most countries have at least some coal in the ground. Widespread availability of coal deposits made it possible for coal to be extracted for local consumption and/or for export to other areas in many locations over the last two centuries, although economically viable reserve are concentrated in five large coal producing countries (China, the U.S., India, Australia, and Russia have 72.4% of all proved recoverable reserves of hard coal in 2015) (BP 2016: 30) (Ciccantell and Gellert 2018). Again, Great Britain and the U.S. could rely on internal coal peripheries and even exported coal, while Japan and China needed more coal, especially metallurgical coal, than their domestic coal peripheries could supply.

In a conceptual sense, natural processes produce coal independent of human processes. This natural product is hugely available and widely spread around the world, but deposits have significant variations in coal quality, the topography of the site of the deposit, and the topographic obstacles to accessing and transporting coal from the deposit. This natural product becomes coal from a human perspective via social action, ranging from building geologic knowledge, conducting exploration, carrying out research and development on extracting and processing coal, creating trade agreements to facilitate coal exports, forming markets, and a variety of other social processes (Ciccantell and Gellert 2018), including the creation of increasingly longer and larger scale coal commodity chains. Increasing the scale and lower the cost of ocean shipping of coal was particularly important for the construction of Japan's and then China's global coal commodity chains.

As the result of this combination of material and social processes, coal reserves increased from 572.7 billion metric tons in 1962 (Brubaker 1967:191) to 1,253,264 million short tons in 2016 (BP 2017), despite global coal extraction of more than 210 billion tons over these 5 + decades. Global proven coal reserves are sufficient for approximately 120 years of production at current levels (IEA 2015: II.27). Any worries about the exhaustion of the earth's coal resources are clearly unfounded; in geologic terms, there are many centuries' worth of coal in the ground. Peak coal is a materially unfounded claim (Ciccantell and Gellert 2018). However, despite coal's geographic abundance and spread, low cost, and geopolitical attractiveness as a domestically available resource, the environmental and social costs of continued reliance on coal are immense.

2.3.2 *Coal Commodity Chains Under British and U.S. Hegemony*

In the context of these material characteristics, the pattern of coal use has evolved significantly over the last three centuries in the capitalist world-economy. Coal's longstanding importance is readily demonstrated by the fact that collecting data on coal extraction began centuries ago. Over the last two centuries, coal production and consumption has grown rapidly in both hegemons and rapidly ascending economies. Coal powered Great Britain's economic ascent to a hegemonic position in the capitalist world-economy from the mid-1700s through the early twentieth century. British industrialization, funded in part by the slave trade that supplied labor for plantation agriculture and mining in the Americas that fed British factories and cities (Williams 1944; Rodney 1982), underlay British military and political power that expanded the British Empire around the globe and overcame repeated economic, political, and military challenges to this hegemony (Bunker and Ciccantell 2005) (Table 2.1).

Coal's contribution to labor productivity and to reduced turnover time and accelerated accumulation of capital since the 1700s depended on coal being available in great volume at low prices. This combination of high volume and low value meant that coal deposits were the primary determinant of early industrial location, with iron and later steel processing plants and factories that consumed iron and steel located near coal deposits (Harris 1988; Isard 1948). Coal extractive peripheries were typically internal to hegemons, core states, and ascending economies. Germany, France, and especially the U.S. followed the British model of fueling their industrialization from domestic coal extractive peripheries. The volume of coal traded internationally was limited, even as recently as the 1960s, as Table 2.2 shows.

In terms of trade flows of coal, from the early 1300s onwards, England was the dominant force in world coal trade (Mitchell 1998a, b). It was not until the late 1800s that this situation changed as the U.S. gained a larger role in this trade and became the leading exporter for much of the post-World War II era (EIA 1983: 3). World coal trade has increased even more rapidly than has world coal extraction and consumption, due to massive increases in the scale of bulk ships that reduced the cost of transporting coal and to declining reserves and production in traditional coal consuming countries in Europe and Japan (Bunker and Ciccantell 2005, 2007), as well as rapid economic growth and coal consumption in China and India in recent years.

These internal coal commodity chains were relatively short, ensured security of supply of this essential raw material and were typically controlled by national private and state-owned firms. Internal coal extractive peripheries from the 1700s through the mid-twentieth centuries relied on labor intensive underground mines with dangerous

Table 2.1 Hard Coal Production in millions of metric tons (anthracite, bituminous, and sub-bituminous). *Source* IEA 2001, 2017 for 1947-present; Mitchell 1998a, b for earlier years; Darmstadter 1971 for 1925, 1938

Year	World	U.K	France	Russia	Germany	U.S	Japan	China
1820		22.3	1.1		1.3	0.3		
1830		30.5	1.9		1.8	0.8		
1840		42.6	3		3.2	2.2		
1850		62.5	4.4		5.3	7.6		
1860		87.9	8.3	0.3	13.6	18.2		
1870		115	13.3	0.7	26.4	37		
1880		149	19.4	3.3	47	72	0.9	
1890		185	26.1	6	70	143	2.6	
1900		229	33.4	16.2	109	245	7.4	
1910		269	38	25.4	153	455	15.7	
1920		233	25	6.7	108	595	29.3	21.3
1925	1184	247	47	15	146	526	32	24.3
1929	1324	262	54	37	177	550	34	25.4
1938	1207	231	47	115	186	355	49	28.8
1940		228	41	140	184	462	56	44.3
1947	1370	204	45	77	71	570	27	14
1950	1435	220	51	95	129	397	39	43
1960	1991	198	56	173	148	392	58	397
1970	2208	147	38	207	118	550	41	354
1973	2237	132	26	221	104	530	25	417
1980	2806	122	20	246	95	710	18	620
1990	3531	93	11	238	77	854	8.3	1051
2000	3633	31	3.8	153	37	895	3	1231
2010	6512	18	0.3	226	15	925	0	3316
2016	6482	4	0	292	4	605	0	3243

Table 2.2 World hard coal trade in millions of metric tons. *Source* IEA 1982, 1992, 2001, 2017, 2019

1960	132
1970	167
1980	263
1990	400
2000	594
2010	1076
2016	1333
2018	1404

working conditions. Miners received low pay and the living conditions of miners, and their families were difficult and unhealthy. Labor unions fought to improve wages and working conditions but were bitterly opposed by coal firms and the state in most cases (Austin and Clark 2012; Freese 2003; Ciccantell 2018; Perdue and Pavela 2012; Ciccantell and Gellert 2018; Gellert and Ciccantell 2020).

During the late 19th and early twentieth centuries, the most important form of globalization and international commodity chains in coal consisted of the networks of coaling stations built by Great Britain and other nations around the world to supply their commercial and military steamships. When possible, core states and firms fomented the development of coal mines near these far-flung coaling stations to reduce the need to and cost of supplying these coaling stations, creating new coal extractive peripheries in a variety of new areas (Barak 2015; Shulman 2015). The widespread geologic availability of coal in Europe and North America made energy imperialism largely unnecessary for coal. However, there were a few key instances in which coal became of focus of imperial strategies. For example, after the development of steamships in Great Britain, Europe, and the U.S. in the mid-1800s, coaling stations to fuel oceangoing steamships and access to relatively nearby coal deposits to supply these coaling stations far from the home country's coal mines became a critical concern for Great Britain, the U.S., and other nations (Barak 2015; Brodie 1941; Gray 2017; Shulman 2015; Ciccantell 2020a, b).

Overall, from the late 1700s through the mid-twentieth century, nation-states followed the British model of creating domestic coal commodity chains linking internal coal peripheries to industrial centers to fuel industrialization and economic ascent. Coal extractive peripheries emerged in dozens of countries, but few of these peripheries captured significant benefits and all suffered significant environmental and social costs.

2.3.3 The Japan-Focused Coal Commodity Chains

Since the 1960s, coal has become one of the most global industries in the world, with 1,373 million tons traded internationally in 2016 (IEA 2017). How did one of the heaviest, bulkiest, lowest value, and most localized industries become so thoroughly transformed in a short period of time into one of the largest and most valuable global commodity chains?

While there were a few exceptions to the old locational rule of reliance on domestic coal peripheries in the first half of the twentieth century, this rule began to change during Japan's reindustrialization based on coal imported from the U.S. and Australia in the 1950s; within two decades, global sourcing of ocean-borne coal supplies made the coastal steel mills in Japan far more competitive than mills in the U.S. and Europe that followed the old locational pattern near coal reserves. For the coal and steel industries, issues of bulk, weight, and transport are the keys to the goal of reducing production costs to make the development of steel and other linked industries globally competitive. The globalization of the coal industry resulted directly from the U.S. led efforts to rebuild Japan after World War II as a geopolitical bulwark in Asia during the Cold War. The U.S. government supported efforts to expand coal production in Australia for export to Japan and helped the Japanese steel firms and the Japanese state create a new model of coastal steel mill and electric power locations that relied on imported metallurgical and steam coal governed by long term contracts, a model that China has replicated in recent years (Bunker and Ciccantell 2007; Hogan 1999a, b; Ciccantell and Gellert 2018; Gellert and Ciccantell 2020).

These geopolitical and economic changes since the mid-twentieth century have dramatically changed the scale and location of coal extraction and use. Huge ocean-going ships carrying 100,000 tons or more of coal at a time began journeys of a thousand to several thousand miles between large ports in Australia, Canada, South Africa, the U.S., and several other coal peripheries to Japan from the 1960s onward at progressively lower transport cost per ton due to economies of scale via these Japan-focused global coal commodity chains. From the 1960s through the early 1990s, coal extractive peripheries enjoyed rising demand and prices, benefitting from what can be termed "hitching your wagon to a rising star" as Japan ascended to become the world's largest steel producer and fastest growing economy in the world during this period (Ciccantell 2009).

One of the most important external coal peripheries created to supply Japan's industrialization in the late twentieth century was located in British Columbia, Canada. Canadian, the U.S. and Japanese firms formed joint ventures to supply coal on the basis of long term contracts, one of the hallmarks of globalization in the late twentieth century. Japanese steel firms and the Japanese state effectively "stole" or redirected a dying coal periphery that had previously supplied the U.S. and Canadian railroads and smelters into a key node in Japan's new global coal commodity chain (Bunker and Ciccantell 2007).

The initial benefits to British Columbia (BC) in western Canada of becoming part of Japan's coal commodity chain in 1960s were profound, as Table 2.3 shows.

The coal industry also had dramatic impacts on local populations and employment in these coal communities. The populations of the coal mining communities grew dramatically from the late 1960s through the early 1980s as the Japan-driven coal boom was underway, as Table 2.4 shows.

Table 2.3 British Columbia coal production and value. *Source* British Columbia Ministry of Energy, Mines and Petroleum Resources: www.empr.gov.bc.ca

Year	Millions of tons	Millions of C$ (current dollars)
1960	0.77	5.6
1965	0.74	5.8
1970	3.2	25.7
1975	9.6	342
1985	22.7	1090
1990	24.6	1001
1991	25	990

Table 2.4 Population of Fernie, Sparwood, and Elkford. *Source* Province of British Columbia, BC Stats: www.bcstats.gov.bc.ca; Statistics Canada. *Note* Sparwood 1961 population is former community of Natal

Year	Population		
	Fernie	Sparwood	Elkford
1961	2661	829	0
1966	2715	1928	0
1971	4422	2990	0
1976	4608	4050	1873
1981	5444	4157	3126
1986	5188	4540	3187
1991	5012	4211	2846

Overall, boom times in the coal industry have produced significant numbers of well-paying jobs extracting and exporting millions of tons of high-quality metallurgical coal from the mountains of southeastern BC. Becoming part of a rapidly ascending economy's global coal commodity chain produced significant benefits for this coal periphery in southeastern BC from the 1960s through the 1980s. The good times from riding the coal "resource rollercoaster", however, would not last forever.

2.3.4 Divergent Paths for Coal and the Rise of China-Focused Coal Commodity Chains in the 2000s

The extraction of coal continued to expand dramatically in the twenty-first century. Despite growing international concern over the environmental unsustainability of fossil fuel use, hard coal production increased by 78% between 2000 and 2016. China became the world's largest coal producer in the 1980s and now produces about half of the world's hard coal, a total of 3.84 billion tons in 2020 (Reuters January 18, 2021). The Chinese coal industry dates back centuries (Wu 2015) and contributed to China's position as the world's largest and most powerful economy until the 1800s, a position to which China seems to be returning (Frank 1998). Coal production and consumption are helping drive China's economic ascent in the twenty-first century, just as they helped maintain China's economic and geopolitical power in earlier eras (Ciccantell and Gellert 2018; Gellert and Ciccantell 2020).

World coal consumption increased by 156% over the past four decades, as Table 2.5 shows, but this is a very uneven process geographically. European consumption peaked in the 1990s, while U.S. consumption peaked in the 2000s and has since begun to decline sharply. The U.S. investment bank Morgan Stanley in 2021 predicted that coal fired power would be ended completely in the U.S. by 2033 (Bloomberg News 2021a). Similarly, Germany has awarded compensation for coal-fired power plant closures of 4.8 GW in December 2020, 1.5 GW in March, and 2.5 GW in April, all to be closed by December 2021 (Franke 2021), reflecting the ongoing and well-funded policy efforts to reduce greenhouse gas emissions there.

Table 2.5 Coal consumption (millions of tons of anthracite, bituminous, sub-bituminous, and lignite). *Source* IEA 2019 Table 2.6 (II.11)

	World	U.S	OECD Europe	China	India
1973	3093	505	1056	414	77
1980	3756	650	1157	626	107
1990	4638	815	1155	1049	220
2000	4748	966	817	1337	357
2010	7135	949	749	3221	683
2016	7455	665	671	3610	914
2018	7,721	614	225	3,745	985

Table 2.6 Coal Imports (millions of tons). *Source* IEA 2001, 2017, 2019

	China	India
1960	0.06	0.01
1970	0	0.004
1980	1.99	0.55
1990	2	5.1
2000	2.1	24.5
2010	184	121
2016	256	200
2018	295	240

However, the ten-fold increase in coal consumption in both China and India since 1973 drove a huge increase in global coal consumption that continues in 2021. While the Chinese government has sought to curtail coal use in recent years because of pollution concerns and promoted rapid growth in renewable energy and nuclear power, it is not clear that coal consumption in China has reached its peak, despite government statements regarding cutting carbon emissions and achieving carbon neutrality by 2060. Perhaps most tellingly regarding divergent futures for coal, "a record-tying 37.8 gigawatts (GW) of coal plants were retired in 2020, led by the U.S. with 11.3 GW and EU27 with 10.1 GW, but those retirements were eclipsed by China's 38.4 GW of new coal plants" (Mining.com April 6, 2021). On the same day that the Chinese government presented its plan for moving to carbon neutrality by 2060, a new US$10 billion project to convert coal to chemical feedstocks in Inner Mongolia was announced (Bloomberg News 2021c). In a further ironic twist, China's worst sandstorm in the past decade was underway in late March and reduced the supply and raised the price of thermal coal whose burning was blamed for the sandstorm and other impacts of climate change (Bloomberg News 2021d).

Coal extraction and demand in India continue to grow as well. Coal India approved 32 coal mining projects totaling US$6.4 billion in early 2021 to bring the company's production to 1 billion tons of coal in 2023, partially replacing the 248.5 million tons of coal imported in 2020 (Bloomberg News 2021b). In 2020, coal produced 65% of India's electricity (Singh 2021).

China has been working to steal Japan's raw materials peripheries and redirect their exports to China, a pattern India is now following (Ciccantell 2009; Nayar 2004). China's hard coal imports remained relatively steady from the 1970s through 2000, but have since exploded, as have India's imports (Ciccantell and Gellert 2018; Gellert and Ciccantell 2020).

China's extremely rapid economic ascent in terms of sustained economic growth rates over the last four decades is well-known (Arrighi 2007). In material energy consumption terms, economic ascent is readily apparent. From only 6.3% of total world energy consumption in 1980, China's share rose to 23.6% in 2018. In metric tons of oil equivalent, Chinese consumption in 2018 was 7.9 times greater than in 1980, while total world consumption had only doubled (BP 2019). This is the essence of economic ascent: a rapidly growing economy that needs more and more resources every year to sustain its growth trajectory at a far higher rate than most other economies and than the world economy as a whole (Ciccantell and Gellert 2018; Gellert and Ciccantell 2020). In a fundamental sense, economic ascent in a rising nation-state restructures elements of the world economy to support this ascent. In some cases, this means opening up new commodity chains that may offer better developmental opportunities for resource peripheries, e.g., Japan's post-World War II ascent gave British Columbia opportunities to earn export revenues, create thousands of jobs, and build new company towns to extract coal in a remote extractive periphery. In other cases, however, a process of ascent such as Germany's rise in the late 19th and early twentieth century can drive imperial expansion and harsh exploitation of populations in extractive peripheries in Africa and ultimately wars to create German hegemony that killed many millions. In recent decades, the ascent of the BRIC nations, particularly India and China, has restructured commodity chains and driven booms in resource prices and employment in many extractive peripheries in coal and other natural resource industries but has also driven deforestation and loss of indigenous lands in the Amazon, exploitation of child labor in mines in Africa, and growing indebtedness of many peripheral and semiperipheral states to China due to the cost of China's Belt and Road Initiative.

The central element of this growth of energy consumption in the late twentieth and early twenty-first centuries in China was coal. Coal consumption provided the largest share of energy to drive Chinese economic ascent (BP 2019). Coal's share did fall during this rapid ascent, but only from 73% to 58.2%, despite extensive efforts by the Chinese government to promote other energy sources, including nuclear power and renewable energy such as hydroelectricity, solar, and wind power. Oil consumption increased by a factor of 7.3 and natural gas consumption by a factor of 19.6 (BP 2019), but coal remains the most important energy sector, due to China's large domestic coal reserves and the creation of a global market for seaborne coal trade by Japan's earlier economic ascent that Chinese steel mills and power companies have been able to utilize to supplement domestic production, particularly with higher quality imported metallurgical coal (Ciccantell and Gellert 2018; Gellert and Ciccantell 2020).

China's large, diverse land area has provided significant quantities of energy resources in addition to coal, most notably large rivers for hydroelectricity, nuclear materials, land for solar installations, and strong sustained winds for wind power generation. Even with these material advantages, rapid economic ascent made imported energy raw materials critical to China's sustained growth (Ciccantell and Gellert 2018). Chinese steel firms are making use of the global seaborne metallurgical coal market created to serve the Japanese steel industry in the 1950s and 1960s (Bunker and Ciccantell 2007) to supply what is now the world's largest steel industry. The Chinese steel industry produces half of all the world's steel every year, providing the essential building block of China's rapid urbanization and infrastructure building and exporting significant quantities to other parts of the world as well (Ciccantell and Gellert 2018).

As was the case of British support for the U.S. economic ascent and the U.S. support for rebuilding Japan, the U.S., and Japan played key roles in China's ascent as the supplier of capital and technology to the rising economy as part of what Arrighi (1994) analyzed as the period of financialization and decline in the existing hegemon and the efforts of financial capital in the hegemon to find new opportunities for investment in rapidly growing economies. Japanese firms played this role in the ascent of China in raw materials, transport, and many other industries in the 1990s and early 2000s (Hogan 1999a; Tse 2000; Todd 1991; Ciccantell and Gellert 2018; Gellert and Ciccantell 2020).

For Canada, Australia, Indonesia, and other coal exporting countries, China's ascent and India's growth and the integration of these coal peripheries into coal commodity chains linked to China and India are increasingly making these extractive peripheries look like successful cases of stealing peripheries from earlier ascendants (Ciccantell 2009). In Canada in particular, China's ascent and creation of a coal commodity chain linking coal peripheries in BC to China's steel industry created a new boom for the coal industry, as Table 2.7 shows.

Again, "hitching your wagon to a rising star" ascendant economy creates significant benefits for a coal extractive periphery, even while the coal industry in the U.S. and Europe declined rapidly. However, wide variations in coal prices and employment even in an apparently "successful" coal periphery in a core nation with much greater negotiating and regulatory power than is the case for peripheral nations during

Table 2.7 BC Coal production, value, and employment. *Source* British Columbia Ministry of Energy, Mines and Petroleum Resources: www.empr.gov.bc.ca

Year	Millions of tons	Millions of C$	Coal employment
1990	24.6	1001	5654
1995	24.4	968	3800
2000	25.7	812	2925
2005	26.7	2300	2754
2010	26	4253	3800
2015	24.4	2000	
2019	26	5522	4460

the twenty-first century makes it clear that today's global coal commodity chains, even when serving the rapidly ascending Chinese economy, continue to leave coal peripheries "riding the resource rollercoaster."

This analysis of coal makes it clear that, over the long term, the key raw material for the first century of the Industrial Revolution is used in far greater volumes today than ever before. More generally, the development of new energy sources is a critical element of the growth of the capitalist world-economy, but new sources do not lead to the abandonment of earlier sources. While the twentieth century is often seen as the Age of Petroleum and oil and natural gas production and consumption rose dramatically, coal remained of critical importance throughout the 20th and into the twenty-first centuries and the Age of Renewables (Gellert and Ciccantell 2020).

2.4 Conclusion: Lessons from the Evolution of Coal Commodity Chains for Extractive Peripheries

This chapter utilized the raw materialist lengthened global commodity chains model (Ciccantell and Smith 2009; Ciccantell and Gellert 2018; Gellert and Ciccantell 2020) to examine the evolution of the coal commodity chain and how the coal global commodity chain has shaped extractive peripheries and their potential for development. Coal was a key ingredient in economic ascent over the past three centuries in Great Britain, the U.S., and Japan (Bunker and Ciccantell 2005, 2007) and remains essential to economic ascent in the twenty-first century for China and India (Ciccantell 2009; Ciccantell and Gellert 2018; Gellert and Ciccantell 2020), despite its contribution to climate change and efforts in many countries to promote a transition toward more sustainable energy systems.

This examination of the evolution of the coal commodity chain over time demonstrated how coal resource peripheries are produced through dynamic and integrated socio-spatial relationships between core and especially ascendant economies and the locations that become their coal extractive peripheries. These socio-spatial relations link core and ascendant economies to coal extractive peripheries, creating coal commodity chains that are essential for core and ascendant economies functioning. These raw materialist lengthened global commodity chains in coal link core and ascendant economies to extractive peripheries, a socio-spatial relationship that creates and reproduces global inequalities and shapes the socioeconomic trajectories of those locations that become coal extractive peripheries. Striking deals with an ascendant economy may offer better terms for a coal extractive periphery because of rapidly growing demand in the rising economy, at least for a period of time, but even in those cases the greatest share of the benefits from the coal commodity chain will accrue to the importing ascendant economy and not to the coal exporting periphery.

This long term view of coal commodity chains also makes it clear that, just as coal did not disappear when petroleum became the key energy source of the twentieth century, coal will not disappear in the twenty-first century, even if renewable energy grows at dramatic rates. The ten-fold increase in global renewable energy consumption (excluding hydroelectricity) between 2000 and 2017 (BP 2019) took place at the same time that global coal consumption was doubling, oil consumption was growing, and natural gas consumption was increasing dramatically. The recent climate change accords may hasten the decline of coal in Europe and the U.S., but the key locations that shape the future of coal will remain China and India. In terms of global environmental sustainability, unless energy use patterns change dramatically in China and India, efforts to address climate change are likely to fail because of the continued essentiality of coal commodity chains for processes of economic ascent and the need for these ascendant economies to continue to create and maintain coal extractive peripheries to support the industrialization and economic ascent of China and India (Ciccantell and Gellert 2018; Gellert and Ciccantell 2020).

For extractive peripheries more generally, the evolution of the coal commodity chain and its impacts on particular nodes in this chain raise a number of questions. For example, given the rapid economic ascent of China and the other BRICs in recent decades, how has the calculus of the costs and benefits of becoming nodes in these nations' new commodity chains changed? As the Chinese government often argues, this era is very different from European colonialism but are extractive peripheries in a better position to capture benefits and avoid costs than they were in earlier eras of European and U.S. ascent? Accelerating climate change, in part driven by the ascent of China and India, is threatening environmental and social sustainability in many extractive peripheries and in the periphery and semiperiphery more generally, e.g., sea level rise in South Asia and the Pacific region that is submerging coastal land and islands, declining availability of fresh water in the Middle East, Africa, and other parts of the world, etc. Does the restructuring of commodity chains and the world economy in the early twenty-first century offer any opportunities for escaping or at least ameliorating the proliferating threats to sustainability?

From an analytic and policy perspective, the raw materialist lengthened GCC model utilized here and the rapidly growing body of literature in critical resource geography offer tools to help us analyze the impacts of resource extraction and use across entire commodity chains and on particular places and peoples in different nodes of these chains. The celebration of the energy transition underway in the wealthy nations of Europe, for example, looks quite different when child labor to extract rare earth minerals enters the analysis. The analysis presented in this chapter highlights how slowly commodity chains can change and the pressing need for critical analyses such as those presented in this volume to inform more just and sustainable futures for extractive peripheries and the world.

References

Arboleda M (2020) Planetary mine: territories of extraction under late capitalism. Verso, London

Arrighi G (1994) The long twentieth century: money, power, and the origins of our times. Verso, London

Arrighi G (2007) Adam Smith in Beijing: lineages of the twenty-first century. Verso, New York
Austin K, Clark B (2012) Tearing down mountains: using spatial and metabolic analysis to investigate the socio-ecological contradictions of coal extraction in Appalachia. Crit Sociol 38(3):437–457
Auty R, Mikesell R (1999) Sustainable development in mineral economies. Clarendon, London
Baglioni E, Campling L (2017) Natural resource industries as global value chains: frontiers, fetishism, labour and the state. Environ Planning A Econ Space 49(11):2437–2456
Bair J (2005) Global capitalism and commodity chains: looking back, going forward. Competition Change 9(2):153–180
Bair J (ed) (2009) Frontiers of commodity chain research. Stanford University Press, Standford, CA
Bair J, Werner M (2011) The place of disarticulations: global commodity production in La Laguna, Mexico. Environ Planning A 43:998–1015
Barak (2015) On outsourcing: energy and empire in the age of coal, 1820–1911. Int J Middle East Stud 47:425–445
Bebbington A et al (2018) Governing extractive industries: politics, histories, ideas. Oxford University Press, Oxford
Bloomberg News (2021a) Coal to exit from US power system by 2033, Morgan Stanley Says. Bloomberg News 1 Feb 2021
Bloomberg News (2021b) Coal India approves 32 mining projects worth $6.4 billion. Bloomberg News 8 March 2021
Bloomberg News (2021c) China's green push kicks off with a new $10bn coal project. Bloomberg News 9 March 2021
Bloomberg News (2021d) China's epic sandstorm lifts the price of coal that caused it. Bloomberg News 24 March 2021
Bridge G (2008) Global production networks and the extractive sector: governing resource-based development. J Econ Geogr 8:389–419
Bridge G, Bradshaw M (2017) Making a global gas market: territoriality and production networks in liquefied natural gas. Econ Geogr 93(3):215–240
British Petroleum (BP) (2016) Various years. BP statistical review of world energy. London.
Brodie B (1941) Sea power in the machine age. Princeton University Press, Princeton, NJ
Brubaker S (1967) Trends in the world Aluminum industry. Johns Hopkins University Press, Baltimore, MD
Bunker SG (1985) Underdeveloping the Amazon. University of Chicago Press, Chicago, IL
Bunker SG, Ciccantell PS (2005) Globalization and the race for resources. Johns Hopkins University Press, Baltimore, MD
Bunker SG, Ciccantell PS (2007) East Asia and the global economy: Japan's ascent, with implications for China's future. Johns Hopkins University Press, Baltimore, MD
Bunker SG (1989) Staples, links and poles in the construction of regional development theories. Sociol Forum 44(4):589–609
Campling L, Havice E (2019) Bringing the environment into GVC analysis: antecedents and advances. In: Ponte S, Gereffi G, Raj-Reichart G (eds) Handbook on global value chains. Edward Elgar, Cheltenham, UK, pp 214–227
Cardoso FH, Faletto E (1969) Dependencia y Desarrollo en America Latina. Siglo Veintiuno, Mexico City
Chase-Dunn C, Hall T (1997) Rise and demise: comparing world systems. Routledge, New York
Ciccantell PS (2018) Chapter 3: mountains, coal, and life in British Columbia and West Virginia. In: Kingsolver A, Balasundarum S (eds) Global mountain regions. Indiana University Press, Bloomington, IN, pp 45–58
Ciccantell PS (2020a) Liquefied natural gas: redefining nature, restructuring geopolitics, returning to the periphery? Am J Econ Sociol 79:265–300

Ciccantell PS, Patten D (2016) The new extractivism, raw materialism, and 21st century mining in Latin America. In: Deonandan K, Dougherty M (eds) Mining in Latin America: critical approaches to the 'New Extractivism' . Routledge, Abingdon, UK, pp 45–62

Ciccantell PS, Gellert PK (2018) Raw materialism and socioeconomic change in the coal industry. In: Debra JD, Matthias G (eds) Oxford handbook of energy and society. Oxford University Press

Ciccantell PS, Gellert PK (2021) Forthcoming 2021. In: O'Hearn D, Ciccantell PS (eds) Migration, racism and labor exploitation in the world-system (39th annual PEWS conference volume). Routledge, New York

Ciccantell PS, Smith DA (2009) Rethinking global commodity chains: integrating extraction, transport and manufacturing. Int J Comp Sociol 50:361–384

Ciccantell PS (2009) China's economic ascent via stealing Japan's raw materials peripheries. In: Hung H-F (ed) China and the transformation of global capitalism. Chapter 6. Johns Hopkins University Press, Baltimore, MD

Ciccantell PS (2020b) Alternatives to energy imperialism: energy and rising economies. J Energy Hist. http://www.energyhistory.eu/en/special-issue/alternatives-energy-imperialism-energy-and-rising-economies

Ciccantell PS (2021) World-systems theory, nature, and resources. In: Himley M, Havice E, Valdivia G (eds) The Routledge handbook of critical resource geography. Routledge, New York

Darmstadter J et al (1971) Energy in the world economy: a statistical review of trends in output, trade, and consumption since 1925. Johns Hopkins University Press, Baltimore

Dinius O, Vergara A (eds) (2011) Company towns in the americas: landscape, power, and working-class communities. University of Georgia Press, Athens, GA

EIA (1983) Various year. Annual energy review. U.S. Energy Information Agency, Washington, DC

Frank AG (1998) ReOrient: global economy in the Asian age. University of California Press, Oakland

Frank AG (1967) Capitalism and underdevelopment in Latin America. Monthly Review Press, New York

Franke A (2021) Germany awards coal closure compensation to Uniper, EPH for 1.5 GW. Platts 1 April 2021

Freese B (2003) Coal: a human history. Penguin Books, London

Freudenburg W (1992) Addictive economies: extractive industries and vulnerable localities in a changing world economy 1. Rural Sociol 57(3):305–332

Freudenburg W, Wilson L (2002) Mining the data: analyzing the economic implications of mining for nonmetropolitan regions. Sociol Inquiry 72(4):549–575. https://doi.org/10.1111/1475-682X.00034

Gaventa J (1980) Power and powerlessness: quiescence and rebellion in an Appalachian valley. Clarendon Press, Oxford

Gellert PK, Ciccantell PS (2020) Coal's persistence in the capitalist world-economy: against teleology in energy 'transition' narratives. Sociol Develop 6(2020):194–221

Gereffi G, Korzeniewicz M (eds) (1994) Commodity chains and global capitalism. Praeger, Westport, CT

Grainger L, Gibson J (1981) Coal utilisation: technology, economics and policy. Graham & Trotman, London

Gray S (2017) Steam power and sea power: coal, the Royal Navy, and the British Empire, c. 1870–1914. Palgrave Macmillan, London

Green H (2010) The company town: the industrial edens and satanic mills that shaped the American economy. Basic Books, New York

Harris JR (1988) The British iron industry 1700–1850. MacMillan Education, Houndmills

Havice E, Campling L (2017) Where chain governance and environmental governance meet: inter-firm strategies in the canned tuna global value chain. Econ Geogr. https://doi.org/10.1080/00130095.2017.1292848

Hayter R et al. (2003) Relocating resource peripheries to the core of economic geography's theorizing: rationale and agenda. Area 35(1):15–23
Hirschman AO (1958) The strategy of economic growth. Yale University Press, New Haven, CT
Hirschman AO (1977) A generalized linkage approach to development with special reference to staples. Econ Develop Cult Change 25
Hogan W (1999a) The steel industry of China: its present status and future potential. Lexington Books, Lanham, MD
Hogan W (1999b) The changing shape of the Chinese industry. New Steel 15(11):28–29
Hopkins T, Wallerstein I (1986) Commodity chains in the world-economy prior to 1800. Review (Fernand Braudel Center): 157–170
Innis H (1956) Essays in Canadian economic history. University of Toronto Press, Toronto
International Energy Agency (IEA) (2015) Various years. Coal information. OECD, Paris
Irarrázaval F, Bustos-Gallardo B (2019) Global salmon networks: unpacking ecological contradictions at the production stage. Econ Geogr 95(2):159–178
Isard W (1948) Some locational factors in the iron and steel industry since the early nineteenth century. J Polit Econ 63(3):203–217
Krannich R et al. (2014) Resource dependency in rural America: continuities and change. In: Jensen L, Ransom E (eds) Rural America in a globalizing world. West Virginia University Press, Morgantown, WV, pp 208–225
Loyola A, Martin MA, Cademortari J (2014) Large mining enterprises and regional development in chile: between the enclave and the cluster. J Econ Geogr 14:73–95
Lucas R (1971) Minetown, milltown, railtown: life in Canadian communities of single industry. University of Toronto Press, Toronto
Malm A (2016) Fossil capital: the rise of steam power and the roots of global warming. Verso, New York
McGraw-Hill (1992) McGraw-Hill encyclopedia of science and technology. McGraw-Hill, New York
Mining.com. (2021) Coal plant development in China offset global retreat. Mining.com 6 April 2021
Mitchell BR (1998a) International historical statistics: the Americas 1750–1993. Macmillan Reference, New York
Mitchell BR (1998b) International historical statistics: Europe 1750–1993. Macmillan Reference, New York
Nayar BR (2004) The geopolitics of China's economic miracle. China Report 40(1):19–47
Perdue RT, Pavela G (2012) addictive economies and coal dependency: methods of extraction and socioeconomic outcomes in West Virginia, 1997–2009. Organ Environ 25(4):368–384
Perloff HS, Wingo L (1961) Natural resource endowment and economic growth. In: Spengler JJ (ed) Natural resources and economic growth 191–212. Resources for the Future, Washington, DC
Perroux F (1955) Note sur la nation de pole de croissance. Economie Appliqué (Jan.–June)
Pomeranz K (2000) The great divergence: Europe, China, and the making of the modern world economy. Princeton University Press, Princeton, NJ
Tse P-K (2000) The mineral industry of China. U.S. Geological Survey, Washington, DC
Reuters (2021) China coal output rises to highest since 2015. Reuters 18 Jan 2021
Rodney W (1982) How Europe underdeveloped Africa. Bogle-L'Ouverture Publications, London, UK
Rostow WW (1960) The stages of economic growth: a non-communist manifesto, 3rd edn. Cambridge University Press, New York
Shulman (2015) Coal & empire: the birth of energy security in industrial America. Johns Hopkins University Press, Baltimore
Singh R (2021) India to double down on coal projects amid climate warnings. Bloomberg News 25 March 2021

Sowers E, Ciccantell PS, Smith DA (2014) Comparing critical capitalist commodity chains in the early twenty-first century: opportunities for and constraints on labor and political movements. J World-Syst Res 20(1):112–139
Sowers E, Ciccantell PS, Smith D (2017) Are transport and raw materials nodes in global commodity chains potential places for worker/movement organization? Labor Soc 20(2):185–205
Sowers E, Ciccantell P, Smith D (2018) Can labor and social movements use commodity chain structure to their advantage? In: Choke points: logistics workers disrupting the global capitalist supply chain. Pluto Press, London
Todd D (1991) Industrial dislocation: the case of global shipbuilding. Routledge, London
Wallerstein I (1974) The modern world system I. Academic Press, New York
Watkins MH (1963) A staple theory of economic growth. Can J Econ Polit Sci 29:141–158
Werner M (2015) Global displacements: the making of uneven development in the Caribbean. Wiley-Blackwell, London
Werner M (2018) Geographies of production I: global production and uneven development. Prog Hum Geogr. https://doi.org/10.1177/0309132518760095
Williams E (1944) Capitalism and slavery. Andre Deutsch, London
Wilson L (2004) Riding the resource roller coaster: understanding socioeconomic differences between mining communities. Rural Sociol 69(2):261–281
Wilson L (2001) The resource roller coaster: social and economic change in two midwestern metal-mining communities. Ph.D. Dissertation, University of Wisconsin-Madison
Wu S (2015) Empires of coal: fueling China's entry into the modern world order, 1860–1920. Stanford University Press, Stanford, CA
Yin R (2017) Case study research and applications: design and methods, 6th edn. Sage, Los Angeles

Paul S. Ciccantell is Professor of Sociology in the Department of Sociology at Western Michigan University and a former Program Officer for the Sociology Program at the National Science Foundation. He received his Ph.D. in Sociology from the University of Wisconsin-Madison. His research examines socioeconomic change over the long term, the evolution of global industries, and the socioeconomic and environmental impacts of global industries, focusing particularly on raw materials extraction and processing and transport industries. He has published books with Johns Hopkins University Press, JAI/Elsevier Press, and Greenwood Press. He has published more than thirty journal articles and book chapters.

Chapter 3
Disarticulations in Resource Peripheries: Bolivia's Oil and Gas Supply Industry

Sören Scholvin

Abstract Research on global production networks (GPNs) has been criticised for its 'inclusionary bias'. The mainstream is focussed on successful development, paying little attention to firms and regions that do not perform well and thus neglecting that the success of some may be tied to the failure of others. This chapter makes a contribution to overcoming the inclusionary bias in order to better understand the prospects of peripheral regions in GPNs. It is inspired by an increasingly vivid academic debate about 'disarticulations'. The author applies this perspective to Bolivia's oil and gas supply industry. He shows that the sector has shifted to a turn-key model, which accounts for the rise of foreign engineering, procurement and construction companies that now capture value at the expense of local suppliers. The latter are downgraded or outright expelled from the corresponding networks. Context factors reinforce the increasingly poor outcomes for Bolivian firms. Many never manage to plug into oil and gas GPNs because of high entry barriers, which even apply to services as generic as catering and transport. However, the chapter also indicates that the disarticulations perspective is itself somewhat biased. It disregards opportunities that integration into the global economy offers in spite of its various downsides.

Keywords Bolivia · Disarticulation · Global production network · Oil and gas · Regional development · Resource periphery

3.1 Introduction

One of the key insights from the work of André Gundar Frank is that the periphery does not suffer from *un*development—that is, a pre-stage of development. On the contrary, *under*development marks all locations beyond the core of the world economy. It results from core–periphery interaction: 'underdevelopment is [...] the product of past and continuing economic and other relations between the satellite underdeveloped and the now developed metropolitan countries' (1969, p. 4). It is 'the

S. Scholvin (✉)
Departamento de Economía, Universidad Católica del Norte, Antofagasta, Chile
e-mail: soren.scholvin@ucn.cl

F. Irarrazaval and M. Arias-Loyola (eds.), *Resource Peripheries in the Global Economy*, Economic Geography, https://doi.org/10.1007/978-3-030-84606-0_3

result of centuries-long participation in the process of world capitalist development' (ibid., p. 6; see also: Frank 1967). This means that the periphery becomes peripheral or, in other words, underdeveloped. It is made peripheral by integration into the world economy.

In the course of the last two decades, mainstream research on the development of peripheral regions has turned away from Frank's work and other basic ideas of world-systems analysis. The global production networks (GPNs) approach has become somewhat dominant in Economic Geography and, to a lesser extent, also in Development Studies. Unfortunately, it does not pay a lot of attention to the fact that not only economic success but also failure of the periphery result at least partly from integration into GPNs. Although Yeung writes that 'strategic coupling [of specific places with global networks] is not automatic and always successful' (2015, p. 6), the articles and books by him and his colleagues almost exclusively deal with successful development. There is little to no recognition that development and prosperity for some people somewhere may induce underdevelopment and misery for other people elsewhere.

To come to a better—a more realistic—understanding of the prospects of the periphery in the global economy, this chapter connects to recent contributions that aim at overcoming the one-sidedness of the GPN literature. It answers a call by Phelps et al. (2018) to engage with 'the dark side of economic geography' by applying the disarticulations concept advanced by Bair and Werner (2011), McGrath (2018), Murphy (2019) and Werner (2016) to Bolivia's oil and gas supply industry. I show how this sector has been affected by downgrading, exclusion and non-participation from/in GPNs. Whereas oil field operators such as Petrobras and Shell as well as their first-tier suppliers—Halliburton, Wintershall and the like—used to engage directly with Bolivian suppliers, contracting them for numerous tasks related to the exploration, extraction and processing of hydrocarbon resources, the lead firms now opt for turn-key projects. Enterprises from Bolivia cannot provide the corresponding financial securities. Foreign engineering, procurement and construction (EPC) companies hence obtain lucrative contracts and then subcontract local suppliers, squeezing the latter's profits. Some local firms have been pushed out of the sector. Others never plug into it under these conditions.

Readers should note that this chapter focuses on Bolivian firms and their prospects. Various certainly important topics are not addressed. This includes, in particular, the resource curse debate (on this issue, see: Hinojosa et al. 2015; Ramírez Cendrero 2014), the impact of hydrocarbon GPNs on indigenous communities and the environment (on this issue, see: Hope 2016; Pellegrini and Ribera Arismendi 2012) as well as political struggles relating to natural resources (on this issue, see: Andreucci 2017; Irarrazaval 2020).

In the following pages, I first summarise academic debates on the dark side of GPNs, which leads me to three forms of disarticulations, different types of intra-regional and inter-regional friction and ruptures, and structural problems that firms from the periphery face in global networks. Next, I explain my research methodologies. I present the empirical findings in the third section. The chapter concludes with a suggestion for follow-up studies and a critique of the disarticulations concept.

3.2 The Dark Side of Global Production Networks

As Ouma et al. (2013) as well as Werner (2016) summarise, research on networks in the global economy—beginning with publications on commodity/value chains and later also including the GPN literature—has shifted from a critical perspective in the tradition of world-systems analysis to one that pays little attention to the uneven distribution of returns among lead firms and suppliers, focussing on mutually beneficial co-ordination of exchange relations instead (esp. Coe et al. 2004; Coe and Yeung 2015; Gereffi and Lee 2012; Kaplinsky and Morris 2016; Morris et al. 2012; Yeung 2009, 2016). Institutions such as the World Bank and World Trade Organisation accordingly praise participation in global networks—guided by sound policies—as the path towards development (Cattaneo et al. 2010; Elms and Low 2013). It seems that if managed wisely, integration into these networks 'boost[s] [economic] growth, create[s] better jobs, and reduce[s] poverty' (World Bank 2020, p. 1).

Indeed, reading the publications by Neil Coe, Gary Gereffi, Henry Yeung and other leading scholars, one hardly learns that regions may stagnate or perform worse because they increasingly participate in the global economy. Admittedly, Gereffi now recognises that 'disinvestment is a common trajectory' (2019, p. 244). Much earlier, he (1994) wrote that few countries have actually been able to copy the East Asian model of development in global networks, with dependency on foreign lead firms often being a liability. Coe and Yeung (2015) mention negative outcomes such as intra-regional differentiation and lock-in to 'structural coupling', which implies a low-value adding position for the corresponding regions.

Werner (2019) rightly notes that this firm-centric literature increasingly acknowledges that there is a dark side of GPNs, but Phelps, Atienza and Arias observe, 'the words ["]uneven development["] do occur frequently in [Coe and Yeung's book] but they remain largely words of an implicit rather than explicit theoretical possibility or an empirical reality' (2018, p. 240). In a paper that deals with 'the dark side of structural coupling', Coe and Hess point out that regions marked by poor labour supply, shortage of venture capital and weak institutions 'are likely to perform very different roles in terms of value capture in global production networks' (2011, p. 132), compared to the typical success cases. It is perhaps revealing that Coe and Hess published these thoughts as a contribution to an edited volume that has received much less attention than the countless articles by them and their colleagues in top journals.

I return to the paper by Coe and Hess later, but for now, it appears fair to say that mainstream research is largely a story about shared benefits of engagement between transnational enterprises, their suppliers and the regions where the latter are located. Regional development results from links with global networks, which lead to distinct trajectories of creation, enhancement and capture of value. These describe how the territories under consideration couple with GPNs in more or less advantageous—but not explicitly harmful—ways (Coe et al. 2004; Coe and Yeung 2015; Yeung 2009, 2016). Even though the term 'periphery' is uncommon in the mainstream GPN literature, it is used on rare occasions: to describe regions that were once peripheral and then succeeded economically because they plugged into global networks. This statement is apparently inspired by modernisation theory. It fully contradicts the observations by Frank that I used as an introduction to this chapter. Relatedly, Bair and Werner (2011) criticise mainstream GPN research for inclusionary bias. It deals with lead firms and their networks, as probably best demonstrated by the ground-laying article by Henderson et al. (2002) and the more recent book by Coe and Yeung (2015). Subordinate suppliers and, more importantly, firms, people and regions that do not participate in GPNs—meaning all those left behind—are of secondary relevance, if addressed at all.

In order to overcome the inclusionary bias, Bair and Werner (2011), McGrath (2018) and Werner (2016) call for a disarticulations perspective (see also: Werner, 2015). Taking up this idea, Murphy explains that 'disarticulations are visible in the exclusions, downgrading processes [and] immiseration [...] that may occur alongside GPN couplings' (2019, p. 950). As his analysis of the tourism industry in Zanzibar (Tanzania) demonstrates, disarticulations comprise three processes. They may serve as a heuristic that structures empirical observations on other business sectors too:

- There is non-participation, as some firms and people never manage to plug into GPNs, sharing the same geographical space with prospering, globalised businesses but lacking opportunities to participate. Zanzibar's foreign-owned luxury resorts host wealthy tourists and generate tremendous profits, while 30% of the local population live below the poverty line and 15% of the local households suffer from severe food insecurity. Non-participation—regarding firms, people and entire regions—has also been observed in the clothing industry in Tunisia (Smith 2015) and India's IT sector (D'Costa 2011), among other cases.
- Exclusion occurs when firms and people are expelled from GPNs. Small indigenous hotels—once frequented by adventurous travellers—are perceived as not trustworthy by the affluent tourists who now visit Zanzibar. Local tour guides rely on walk-in clients, but they are increasingly bypassed by full-package operators from abroad. Others show that exclusion from GPNs often occurs as disinvestment, with export-oriented industries relocating in pursuit of cheap labour and advantageous international trade regimes (Álvarez Medina and Carrillo 2014; Pickles and Smith 2011; Werner 2015).
- Locals sometimes remain within GPNs but suffer from downgrading. Indigenous enterprises in Zanzibar have shifted to employing drivers, guides and other staff on an as-needed basis, thus contributing to the misery of local labour (for other cases of workforce stratification, see: Alford et al. 2017; Godfrey 2015; Rossi 2013). Generally speaking, if local firms manage to upgrade to more sophisticated tasks, this does not necessarily mean higher wages for their employees or broad-based regional development (Dussel Peters 2008; Palpacuer 2008).

In a similar vein, Kaplinsky (2000) observes that the periphery suffers from what he calls immiserating growth. For example, while indigenous firms in Brazil's shoe industry and the textile industry in the Dominican Republic increased their exports in the course of the 1990s, salaries declined (if measured in international purchasing

Table 3.1 The Dark Side of Strategic Coupling. *Source* Author's own draft based on Coe and Hess (2011)

	Inter-regional	Intra-regional
Rupture	Disinvestment	Displacements, evictions, etc
Friction	Unequal value capture	Uneven resource allocation

parity). Kaplinsky argues that the reason for this is that the periphery suffers from entry barriers, being kept in a low-value adding position, which is marked by less and less value capture due to increased worldwide competition. Gibbon and Ponte (2005) go a step further. They suggest that the periphery must downgrade voluntarily to remain within GPNs. Further to that, value increasingly results from intangible assets such as not only intellectual property but also access to finance and the capacity to co-ordinate global networks, with the latter two being typical strengths of enterprises from the core (Kaplinsky 2000).

As noted, the scholars who have shaped GPN research are not ignorant of downsides and pitfalls. Coe and Hess (2011) draft a four-field matrix in this regard, shown in Table 3.1. First, there is disinvestment, meaning an inter-regional rupture. Non-local firms leave the region under consideration and/or local companies lose non-local markets. Second and if incorporation into GPNs persists, value capture may be highly unequal between that region and its non-local GPN partners, leading to inter-regional friction. Third, displacements, evictions and similar processes—usually accompanied by political exclusion and the cutting of economic ties—mark ruptures within a region. Fourth, there may be intra-regional friction because of uneven resource allocation against the backdrop of various social cleavages, with some people benefitting from the region's integration into GPNs, whereas others do not (or hardly so).

Quite surprisingly, this engagement with the dark side of global networks leads to policy recommendations that are everything but critical: Kaplinsky suggests that companies from the periphery become more efficient, link up with partners, and provide new and higher value-adding products and services. Coe and Hess propose for regional institutions to work harder, better moulding regional assets to facilitate development. One wonders whether these suggestions are feasible given the just mentioned challenges.

Bair, McGrath, Werner and probably also Murphy have something else in mind. For them, the downgrading or outright exclusion of some is necessary for the upgrading and inclusion of others. Fortune begets misery. McGrath accordingly emphasises that 'the dis/articulations perspective […] questions the [euphemistic] trajectories of capitalist development' (2019, p. 9). Werner rhetorically asks, 'do firms or regions locked in to "low-value" […] functions make it possible for other firms or regions to be linked in more "value adding" ways?' (2019, p. 951). In other words, research that aims at overcoming the inclusionary bias must address how 'rendering people and places as exterior to global production networks is part and parcel of on-going accumulation through these arrangements and the remaking of uneven development that they entail' (Werner 2016, p. 458). In the section after next, I show how this idea applies to Bolivia's oil and gas supply industry.

3.3 Research Methodologies

My analysis is based on 16 open-ended, narrative interviews with representatives of business organisations, public authorities and, most importantly, local and non-local enterprises. These are listed in Table 3.2. The interviews were conducted in the city of Santa Cruz. The objective was to learn about inter-firm relations—especially with regard to local suppliers—as well as challenges and opportunities for local companies. I identified the interviewees via LinkedIn, by snowball sampling and was put in contact with some by the Bolivian Chamber of Hydrocarbons. A guideline of eight questions was used and slightly adapted before each interview, reflecting on the interviewee's area of expertise and the exact nature of his/her company or organisation. I recorded 13 interviews and later structured them along with the three aforementioned processes of disarticulations. In three cases, I could only take notes. Most interviewees did not speak as official representatives of their company or organisation, but I provide the corresponding information for better contextualisation. All direct quotes are my own translation.

Further to that, I obtained basic information from the website 'A Barrel Full'. I refer to Bolivian newspapers and reports by the Fundación Jubileo and the Fundación

Table 3.2 Interviews. *Source* Author's own compilation

ID	Professional affiliation	Date
1	Oil major	31 July 2017
2	Consultant	31 July 2017
3	Business association	1 August 2017
4	Local supplier	1 August 2017
5	Oil major	1 August 2017
6	Oil major	3 August 2017
7	Oil major	3 August 2017
8	Local supplier	3 August 2017
9	Local supplier	3 August 2017
10	Local supplier	4 August 2017
11	National oil company	4 August 2017
12	Public authorities of the province	4 August 2017
13	Consultant	7 August 2017
14	Broker	8 August 2017
15	Oil major	8 August 2017
16	Foreign service provider	8 August 2017

Milenio—two civil society organisations that publish on economic, political and social topics. Unfortunately, most available datasets and publications are limited to the volume of oil and gas exports, investment in new exploration, royalties and taxes that derive from the sector as well as its absolute and relative decline/growth at the subnational level. To my best knowledge, they provide no substantial information on the local supply industry. Given that the empirical analysis, therefore, mainly rests on interviews, triangulated with information from the just mentioned sources as far as possible, it has certain strengths and weaknesses that are typical of qualitative research: it is rather strong in explaining causal mechanisms, whereas there are shortcomings in terms of generalising statements based on individual interviews.

3.4 Downgrading, Exclusion and Non-participation in Bolivia

Bolivia's massive natural gas deposits were discovered in the late 1990s, but the first oil findings date back to the 1920s, made in close proximity to the city of Santa Cruz. Ever since then, the sector—in terms of oil majors as well as domestic and foreign suppliers—has concentrated there, although the largest active fields are now located about 600 km south, in the province of Tarija (A Barrel Full 2015; see also: Map 3.1). The only exception from this spatial pattern results from the re-nationalisation of Bolivia's natural resources in 2006, which has boosted the influence of the state and the state-owned oil company YPFB—the sole downstream operator and a joint venture partner in all upstream projects.[1] National ministries and the headquarters of YPFB are in La Paz, where lead firms have small offices and indigenous suppliers rely on acquaintances for handling administrative matters (Interview 1, 4, 7 and 15).

An interviewee who works for an oil major explained that due to this track record, there are local firms in Santa Cruz 'with lots of years of experience'. They have the technical know-how needed to carry out specialised services such as the construction of pipelines and even processing facilities for natural gas. These firms have 'experienced technicians' and 'highly skilled staff' (Interview 1). Such location advantages are reinforced by the fact that Santa Cruz is the economic hub of Bolivia beyond oil and gas too, meaning that it offers a relatively well-developed industrial base and good transport infrastructure that connects domestically and internationally (Interview 11; see also: Fundación Milenio 2018b). In other words, it appears that conditions are favourable for Santa Cruz and firms from there to plug into oil and gas GPNs, and benefit considerably.

[1] The oil and gas sector consists of down-, mid- and upstream activities. Upstream includes searching for oil and gas fields, drilling wells and operating these wells. Midstream involves transport, storage and wholesale marketing of crude and purified/refined products. Downstream comprises refining crude oil and purifying raw natural gas and marketing and distribution of consumer products.

3.4.1 Downgrading

Besides the drop of the oil price in 2014, the most widely discussed challenge to Bolivia's hydrocarbon sector is the lack of exploration. Since resources are continuously extracted, a dynamic and prospering oil and gas industry depends on new deposits being found and tapped. If exploration does not keep pace with extraction, the industry will virtually run dry. In Bolivia, merely 70 wells were drilled from the re-nationalisation of YPFB in 2006 until 2019. Half of them did not lead to any findings (Mining Press 2020b). The country's natural gas production is, therefore, expected to decrease from its maximum of 61 million cubic metres a day in 2014 to 34 million in 2025—hardly enough to keep exporting (Mining Press 2020a; see also: Fundación Milenio 2018a). It is obvious that such a scenario has severe implications for local firms that participate in hydrocarbon GPNs. In the following sub-sections, I however apply the disarticulations heuristic to elaborate on challenges to regional development that result from features of global networks, not from declining resource extraction.

Reality looks different though. A local businessman explained that his company used to be contracted for construction work by oil majors—Repsol, Total and others—and specialised service providers such as Baker Hughes and Schlumberger. He described these relations as 'partnerships'. Nowadays, his former clients only contract EPC companies, which are usually from Spain or the United States. The EPC companies then subcontract Bolivian suppliers—like the interviewee's firm—and put high pressure on them to reduce costs (Interview 10). Another interviewee concurred, saying that in the past, his company provided small and medium-sized modules for larger projects, collaborating with other Bolivian firms and being contracted directly by oil majors—'a model that worked [for everyone]', as he put it. Today, 70% of his business is with EPC companies, which the interviewee called 'a new monster' (Interview 8).

Why have EPC companies emerged? And why do they have such a bad reputation? Being convinced to work more cost-efficiently this way, foreign oil majors and YPFB have decided to switch to turn-key projects instead of hiring numerous suppliers for specific tasks. This means that there is only one direct contractor per project and that firm has to provide a complete package. The problem with turn-key projects is that these projects are so large that their execution requires heavy insurances; to cover the potential costs of delays or environmental damage, for instance. Bolivian firms do not have enough money to pay for such insurances (Interview 3, 9, 10). An executive of an oil major confirmed the high relevance of financial capacities: 'before formally inviting [a company to bid for a project], we check whether they have a sufficient financial backbone' (Interview 15).

This entry barrier is reinforced by Bolivia's weak domestic banking sector (Interview 8). A consultant with prior experience in YPFB and an EPC company explained that '[financial] guarantees for a decent project are about 7% [of the total price]. So if you built a large plant and it costs [USD] 400 million, 7% is [USD] 28 million. No local company [can obtain that amount of money], primarily because of the conditions that banks impose. Banks [here in Bolivia] ask you for 1:1 guarantees [...] That's where foreign companies come in: they bring the financial capacities' (Interview 2).

This restructuring of oil and gas GPNs does not, in most cases, mean that indigenous suppliers are completely expelled. Following the new model, 'these big [EPC] companies [...] bring the money. They have know-how [in engineering and construction], of course, but what they do [is]: they come to Bolivia and subcontract local firms. So in fact, the [physical] work is done by Bolivian firms', which are downgraded from contractors to subcontractors and, as noted, see their profits squeezed. Usually, there is no input by the EPC companies beyond finance and contacts to the sector's lead firms: 'the know-how is here [...] All work is done by Bolivian engineers. The foreign enterprises just sent supervisors' (Interview 2).[2] Hence, the EPC companies are 'construction firms that don't construct [anything]', as the aforementioned businessman put it (Interview 10).

For providing financial security, contacts to lead firms and co-ordinating the subcontracting of many Bolivian suppliers, the EPC companies apparently add 5–10% to the actual costs of each project, which then becomes their own profit (Interview 2). In other words, the aforementioned large plant worth USD 400 million means revenues of USD 20–40 million for the respective EPC company. Local suppliers, meanwhile, carry all risks for delays, malpractices and the like (Interview 8, 9). Oil majors of course demand that the EPC companies have adequate insurances. As said, the need for insurance is the very reason for the rise of EPC companies. Yet, the EPC companies impose the same requirements—cut into smaller pieces that reflect the corresponding tasks—on the Bolivian subcontractors (Interview 14). If anything goes wrong, the EPC companies will be liable vis-à-vis the oil majors, but they pass this liability to indigenous firms.

3.4.2 *Exclusion*

The turn-key model, which accounts for downgrading, apparently also means that Bolivian suppliers are now excluded from some segments of oil and gas GPNs to which they used to have access. Executives from an oil major pointed out that indigenous firms do not have sufficient machinery and technology for large-scale turn-key projects, saying that 'the idea of local companies substituting [foreign service providers] is fiction' (Interview 7), at least unless lead firms are willing to return to directly contracting numerous firms for more limited, specific tasks. Another interviewee reasoned that with regard to the largest turn-key projects,'there are no local companies that have ever built anything of that size [on their own]. Normally, you

[2] Foreign service providers usually have a few ex-pats in Bolivia and contract much more local staff for less specialised tasks. The share of Bolivian employees is set by the law at a minimum of 90%, but it is usually higher for cost reasons (Interview 5).

do that with a foreign firm because first, experience and second, capital' (Interview 11).

Interviewees from a business association were also sceptical regarding the capacities of local firms or, rather, corresponding over-estimations: 'Bolivian companies will tell you that they can do it, [but] the economic capacity of a Bolivian company, even if it's a group of companies, considering the guarantees they can provide [...] you realise that the conditions are just not right and [that's when] we are talking about projects of little technology, little know-how but major risks' (Interview 3). Indeed, as a consequence of miscalculation of costs and risks of increasingly large projects (but also due to downgrading), many local firms have gone bankrupt (Interview 9). Another challenge results from the fact that EPC companies provide all construction plans and materials needed for the corresponding projects, thus excluding Bolivian suppliers that used to play the corresponding roles (Interview 1, 9).

There also appears to be corruption or at least obscure personal relations that lead to the exclusion of local companies. An interviewee reasoned that since re-nationalisation, 'the government controls everything [and] the EPC companies meet with the government. They tie themselves to it' (Interview 9). Against this backdrop, it is unlikely that calls by the provincial government of Santa Cruz to tighten legislation on local content—minimum prices for certain services that can be provided locally, for example (Interview 12)—will be heard (regardless of whether such interference with the economy does more bad or good). It is obviously difficult to find evidence to back up such claims. However, YPFB has an impressive track record of corruption and illicit business dealings. Seven of the eight presidents of the company during the 14 years of Morales's presidency were accused of corruption. One of them—Santos Ramírez—was condemned to 12 years imprisonment (El Día 2020a). YPFB had three presidents under the interim centre-right government—in power since Morales was forced into exile in late 2019 until the end of the following year—and one of them was a fugitive from justice at the time of writing this chapter (El Mundo 2020). Further to that, there are now accusations that YPFB had secret accounts during the Morales era, with millions of US dollars disappearing (El Día 2020b; Jornada 2020).

Kaup (2020) relatedly reports that party affiliates with little to no experience in hydrocarbons were appointed to the upper echelons of YPFB under Morales. They then willingly implemented a political agenda or were forced to do so, even though this agenda may not have been rational from a business point of view.

Yet, it is important to note that exclusion pre-dates the rise of the EPC companies. A broker explained that with the privatisation of YPFB in 1996, foreign companies established insurances as a principle for the sector. In his experience, Bolivian firms need about USD 1 million as collateral for an average project with YPFB (i.e. a task of limited scope that must be coordinated with other suppliers to complete the entire project). Foreign clients demand insurances that cover twice as much. The interviewee added that 'these are costs that hurt [local firms]', decreasing their competitiveness vis-à-vis foreign service providers (Interview 14). Thus, even if foreign oil majors and YPFB were willing to return to directly contracting numerous local suppliers, many Bolivian firms would probably remain excluded from the sector.

3.4.3 Non-participation

Whereas downgrading describes how firms move to a less lucrative position in GPNs and exclusion refers to companies that used to be part of such networks but are then expelled, non-participation is about local players that never integrate into the GPNs under consideration. It also applies to segments of these GPNs that never see indigenous participation. In addition to finance, the high standards of the oil and gas sector matter as entry barriers. For the most sophisticated tasks related to the exploration and extraction of hydrocarbons, there are no Bolivian suppliers (Interview 7). This is not surprising, considering that such services are provided by less than a dozen firms worldwide: Baker Hughes, Halliburton, Schlumberger, Wintershall and a few others.[3] There is also a shortage of perforation teams in Bolivia. These usually come from Argentina and Brazil (Interview 5). An interviewee from an oil major added that many inputs—tubes for pipelines, for instance—are simply not available on the domestic market (Interview 15).

Even with regard to generic services, there are important challenges. An interviewee who works for a transnational service provider explained that her company locally subcontracts services such as office cleaning, security and transport. Subcontracting in transport services—small aircraft, helicopters, lorries and other means of road transport—is particularly difficult because not all freight forwarders are willing to participate in obligatory training and invest in vehicles that meet her company's standards. Providing an example that demonstrates that not everyone can supply services in the hydrocarbon sector, she said that her colleagues who travel to distant oil and gas fields must not use public transport, which is informal in Bolivia, because it does not comply with her company's security regulations (Interview 16). Other interviewees added that those who provide catering for camps in disparate locations ought to meet the same standards as caterers of international airlines (Interview 6).

If such barriers are already challenging for firms from Santa Cruz, they will be almost insurmountable for their peers from more peripheral locations. As noted, Bolivia's largest active oil and gas fields are in the province of Tarija. Interviewees said that they 'would very much like to contract service in Tarija, but there are just no capacities there' (Interview 7).

Unlike many other sectors, oil and gas have a formalised entry barrier that reflects the high relevance of standards. Before a company can provide products or services to an oil field operator, its strategic partners and specialised suppliers, it has to enter a list of approved vendors. In Bolivia, the law mandates that oil field operators continuously update the vendors' list in collaboration with YPFB. The list is not publicly available, but according to information obtained by the Fundación Jubileo (2016), it contained only 103 firms in 2011. Seventy-seven of them were Bolivian. About 90% of the Bolivian firms had their head office in Santa Cruz, with the vast majority providing products and services in engineering. Only four catering companies and two freight

[3] As Bridge (2008) explains, the sharp division between globally active oil field service providers and local suppliers, which concentrate on activities that are not overly sophisticated or, in many cases, generic, is characteristic of the upstream sector.

forwarders for heavy equipment and machinery were on the list, indicating that only a marginal share of Bolivian enterprises is eligible for oil and gas-related projects.

The rise of EPC companies reinforces such problems. Almost all interviewees from oil majors emphasised that they are committed to upskilling local suppliers, providing training on bookkeeping, protection of the environment as well as quality and safety standards (Interview 1, 6, 7, 15). However, EPC companies follow a short-term business model. They carry out projects on demand on an in-and-out basis but do not invest in Bolivia (Interview 2).

As a side note, it did not, certainly, help that the Morales government understood hydrocarbon-based development as in-country processing of extracted resources. A report by the Fundación Jubileo (2017a) and an article by Ramírez Cendrero (2014) summarise that YPFB and a newly formed hydrocarbon industrialisation company—EBIH, by its Spanish acronym—were instructed to build several petrochemical complexes (see also: Map 3.1). In recent years, some of these projects have been cancelled. A few new ones have been conceived. As a general tendency, project implementation has been slow and it appears that there is neither an economic case nor a sufficient resource base for many of them because of the aforementioned lack of exploration (Fundación Jubileo 2017b). Even worse, EBIH and YPFB contracted EPC companies for the construction, thus reinforcing downgrading, exclusion and non-participation of indigenous suppliers. The development of a supply industry, meanwhile, is not even mentioned in the government's hydrocarbon strategy, although this document is more than 500 pages long (Ministerio de Hidrocarburos y Energía 2008).

3.5 Conclusion

The chapter applied the disarticulations concept to Bolivia's oil and gas supply industry. I showed that hydrocarbon GPNs have shifted to a turn-key model, with EPC companies from abroad rising against the backdrop of their superior financial capacities and capturing much value at the expense of indigenous firms. The latter are downgraded, now being subcontracted instead of interacting directly with lead firms. In other words, entry barriers that result from access to finance and the capacities to co-ordinate networks—as highlighted by Kaplinsky (2000)—account for the dire situation of Bolivian enterprises, which also run an increasing risk of exclusion from GPNs for economic and political reasons. Because of high standards, even generic service provision is not easy. In consequence, only a marginal share of Bolivia's indigenous enterprises manage to plug into the hydrocarbon sector. The vast majority do not participate at all. This problem is reinforced by the absence of a national policy to support the supply industry and the fact that the EPC companies are not committed to upskilling suppliers.

In follow-up studies, the underdevelopment of other provinces merits attention. Referring to the analytical framework proposed by Coe and Hess (2011), the Bolivian oil and gas supply industry suffers from inter-regional friction, meaning unequal value capture to the disadvantage of local players. Inter-regional ruptures do not apply. Friction and ruptures at the intra-regional level were not investigated here, but it appears likely that firms based in Santa Cruz—while suffering from subordination to foreign lead firms and EPC companies—squeeze value out of smaller businesses and labour in more peripheral parts of the country, for example in Tarija. On the global scale, Bolivia—and therein Santa Cruz—is peripheral and thus underdeveloped. Yet on the national scale, Santa Cruz is the core and presumably drives underdevelopment elsewhere in the country. Slightly rephrasing what Frank eloquently wrote, 'a photograph of the world taken at a point in time [...] consists of a world metropolis [...] and international satellites like [Santa Cruz]. Since [Santa Cruz] is a national metropolis on its own right, the model consists further of its satellites: the provincial metropolises [...] and their regional and local satellites in turn' (1967, p. 146).

On another matter, I find the disarticulation concepts somewhat biased and thus potentially misleading, despite the significant insights gained. Reading the articles by Bair, McGrath and Werner (but less so Murphy's contribution), it is hard to miss that these scholars are driven by anti-capitalist convictions. They rightly criticise the mainstream GPN literature for the inclusionary bias, but one must not overlook that participation in GPNs does offer opportunities. With regard to Bolivia, a consultant explained that many local firms began as suppliers of materials needed by foreign companies. This has enabled them to gain new skills and eventually venture into construction and maintenance (Interview 13). Directors and leading executives of indigenous suppliers pointed out that they have expanded into the neighbouring countries, usually being contracted by oil majors for whom they previously worked in Bolivia (Interview 8, 9, 10).

I cannot offer a solution for the apparent dissonance between the concepts of disarticulation and underdevelopment, on one side, and my own empirical observations, on the other. One avenue is to argue that even though the just mentioned opportunities exist, this chapter demonstrated that there are major pitfalls of participation in the global economy and these prevent a structural transformation of the regional economy of Santa Cruz or, even worse, lead to a reconfiguration marked by downgrading, exclusion and non-participation. Another line of reasoning is that research in the tradition of world-system analysis is about 'big D' development (or, rather, the impossibility thereof), whereas the GPN literature deals with 'little d' development, meaning the creation of individual losers and winners on account of unfolding economic change.

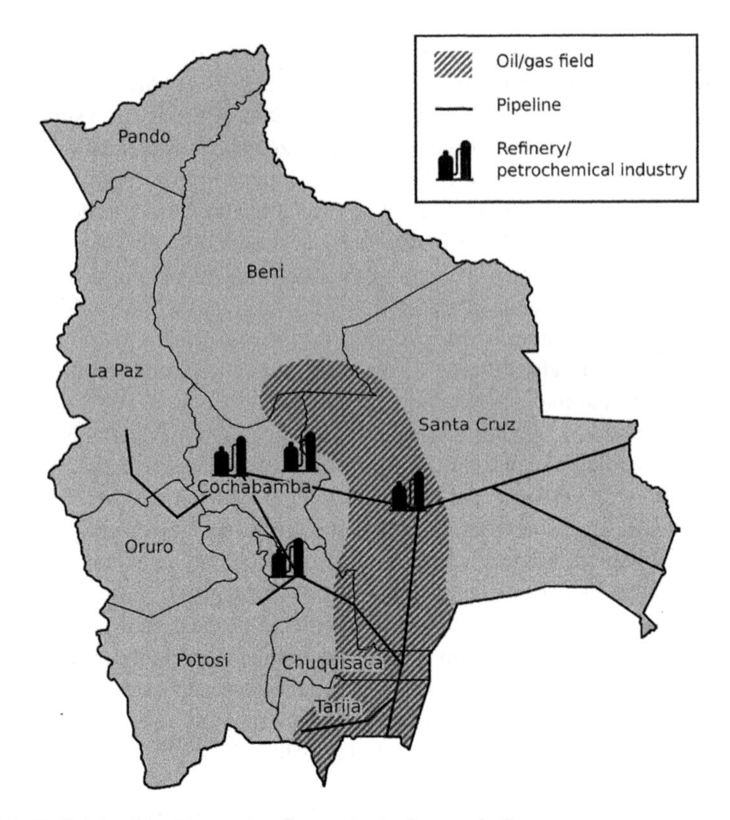

Map 3.1 Bolivia's oil and gas sector. *Source* Author's own draft

References

A Barrel Full (2015) Bolivia oil and gas profile. http://abarrelfull.wikidot.com/bolivia-oil-and-gas-profile

Alford M, Barrientos S, Visser M (2017) Multi-scalar labour agency in global production networks: contestation and crisis in the South African fruit sector. Dev Chang 48(4):721–745

Álvarez Medina L, Carrillo J (2014) Restructuring of the automotive industry in the North American Free Trade Agreement (NAFTA) region from 2007 to 2011. Int Rev Manage Bus Res 3(4):2120–2130

Andreucci D (2017) Resources, regulation and the state: struggles over gas extraction and passive revolution in Evo Morales's Bolivia. Polit Geogr 61:170–180

Bair J, Werner M (2011) Commodity chains and the uneven geographies of global capitalism: a disarticulations perspective. Environ Plan A 43(5):988–997

Bridge G (2008) Global production networks and the extractive sector: governing resource-based development. J Econ Geogr 8(3):389–419

Cattaneo O, Gereffi G, Staritz C (2010) Global value chains in a post-crisis world: a development perspective. World Bank

Coe NM, Hess M, Yeung HW, Dicken P, Henderson J (2004) 'Globalizing' regional development: a global production networks perspective. Trans Inst Br Geogr 29(4):468–484

Coe NM, Hess M (2011) Local and regional development: a global production network approach. In: Pike A, Rodríguez Pose A, Tomaney J (eds) Handbook of local and regional development, pp 128–138. Routledge
Coe NM, Yeung HW (2015) Global production networks: theorizing economic development in an interconnected world. Oxford University Press
D'Costa AP (2011) Geography, uneven development and distributive justice: the political economy of IT growth in India. Camb J Reg Econ Soc 4(2):237–251
Dussel Peters E (2008) GCCs and development: a conceptual and empirical review. Compet Chang 12(1):11–27
El Día (2020a) 11 cabezas de YPFB marcados por interinatos y corrupción. https://www.eldia.com.bo/index.php?cat=426&pla=3&id_articulo=304397
El Día (2020b) Hidrocarburos denuncia depósitos irregulares de dinero de YPFB a cuentas particulares en el gobierno de Evo. https://www.eldia.com.bo/index.php?cat=426&pla=3&id_articulo=309937
El Mundo (2020) Ordenan aprehensión del expresidente de YPFB. https://elmundo.com.bo/2020/06/19/ordenan-aprehension-del-expresidente-de-ypfb
Elms DK, Low P (eds) (2013) Global value chains in a changing world. WTO
Frank AG (1967) Capitalism and underdevelopment in Latin America: historical studies of Chile and Brazil. Monthly Review Press
Frank AG (1969) Latin America—underdevelopment or revolution: essays on the development of underdevelopment and the immediate enemy. Monthly Review Press
Fundación Jubileo (2016) Impacto local generado por el sector hidrocarburos. https://jubileobolivia.com/Publicaciones/Revistas-Especializadas/impacto-local-generado-por-el-sector-hidrocarburos
Fundación Jubileo (2017a) Industrialización de los hidrocarburos. https://jubileobolivia.com/Publicaciones/Documentos/industrializacion-de-los-hidrocarburos
Fundación Jubileo (2017b) Qué pasó con la estrategia boliviana de hidrocarburos? https://jubileobolivia.com/publicaciones/Revistas-Especializadas/iQue-paso-con-la-estrategia-boliviana-de-hidrocarburos
Fundación Milenio (2018a) El cuadro crítico de la producción y reservas de gas natural. https://fundacion-milenio.org/coy-389-el-cuadro-critico-de-la-produccion-y-reservas-de-gas-natural
Fundación Milenio (2018b) Santa Cruz es la región que aporta al PIB del país con el 28,7%. https://fundacion-milenio.org/el-dia-santa-cruz-es-la-region-que-aporta-al-pib-del-pais-con-el-287
Gereffi G (1994) The organization of buyer-driven global commodity chains: how U.S. retailers shape overseas production networks. In: Gereffi G, Korzeniewicz M (eds) Commodity chains and global capitalism, pp 95–122. Praeger
Gereffi G (2019) Economic upgrading in global value chains. In: Ponte S, Gereffi G, Raj-Reichert G (eds) Handbook on global value chains, pp 240–254. Elgar
Gereffi G, Lee J (2012) Why the world suddenly cares about global supply chains. J Supply Chain Manag 48(3):24–32
Gibbon P, Ponte S (2005) Trading down: Africa, value chains, and the global economy. Temple University Press
Godfrey S (2015) Global, regional and domestic apparel value chains in Southern Africa: social upgrading for some and downgrading for others. Camb J Reg Econ Soc 8(3):1491–1504
Hinojosa L, Bebbington A, Cortez G, Chumacero JP, Humphreys Bebbington D, Hennermann K (2015) Gas and development: rural territorial dynamics in Tarija, Bolivia. World Dev 73:105–117
Hope J (2016) Losing ground?: extractive-led development versus environmentalism in the Isiboro Secure Indigenous Territory and National Park (TIPNIS). Bolivia. Extr Ind Soc 3(4):922–929
Irarrazaval F (2020) Contesting uneven development: the political geography of natural gas rents in Peru and Bolivia. Polit Geogr. https://doi.org/10.1016/j.polgeo.2020.102161
Jornada (2020) Bs 100 millones fueron retirados de YPFB Refinación y depositados en cuentas privadas. https://jornada.com.bo/bs-100-millones-fueron-retirados-de-ypfb-refinacion-y-depositados-en-cuentas-privadas

Kaplinsky R (2000) Globalisation and unequalisation: what can be learned from value chain analysis? J Dev Stud 37(2):117–146
Kaplinsky R, Morris M (2016) Thinning and thickening: productive sector policies in the era of global value chains. Eur J Dev Res 28(4):625–645
Kaup BZ (2010) A neoliberal nationalization?: the constraints on natural-gas-led development in Bolivia. Lat Am Perspect 37(3):123–138
McGrath S (2018) Dis/articulations and the interrogation of development in GPN research. Prog Hum Geogr 42(4):509–528
Mining Press (2020a, 10 April) Expertos: en cinco años Bolivia no tendrá gas para exportar. http://miningpress.com/nota/328111/expertos-en-cinco-anos-bolivia-no-tendra-gas-para-exportar
Mining Press (2020b, 4 October). Milenio: crisis de los hidrocarburos en Bolivia. http://miningpress.com/nota/332610/milenio-crisis-de-los-hidrocarburos-en-bolivia-el-informe'
Ministerio de Hidrocarburos y Energía [of Bolivia] (2008) Estrategia boliviana de hidrocarburos. https://www.hidrocarburos.gob.bo/phocadownload/Estrategia%20Boliviana%20de%20Hidrocarburos%202008.pdf
Morris M, Kaplinsky R, Kaplan D (2012) One thing leads to another: promoting industrialisation by making the most of the commodity boom in sub-Saharan Africa. Lulu
Murphy JT (2019) Global production network dis/articulations in Zanzibar: practices and conjunctures of exclusionary development in the tourism industry. J Econ Geogr 19(4):943–971
Ouma S, Boeckler M, Lindner P (2013) Extending the margins of marketization: frontier regions and the making of agro-export markets in northern Ghana. Geoforum 48:225–235
Palpacuer F (2008) Bringing the social context back in: governance and wealth distribution in global commodity chains. Econ Soc 37(3):393–419
Pellegrini L, Ribera Arismendi M (2012) Consultation, compensation and extraction in Bolivia after the 'left turn': the case of oil exploration in the north of La Paz department. J Lat Am Geogr 11(2):103–120
Phelps NA, Atienza M, Arias M (2018) An invitation to the dark side of economic geography. Environ Plan A 50(1):236–244
Pickles J, Smith A (2011) Delocalization and persistence in the European clothing industry: the reconfiguration of trade and production networks. Reg Stud 45(2):167–185
Ramírez Cendrero JM (2014) Has Bolivia's 2006–12 gas policy been useful to combat the resource curse? Resour Policy 41:113–123
Rossi A (2013) Does economic upgrading lead to social upgrading in global production networks?: evidence from Morocco. World Dev 46:223–233
Smith A (2015) Economic (in) security and global value chains: the dynamics of industrial and trade integration in the Euro-Mediterranean macro-region. Camb J Reg Econ Soc 8(3):439–458
Werner M (2015) Global displacements: the making of uneven development in the Caribbean. Wiley
Werner M (2016) Global production networks and uneven development: exploring geographies of devaluation, disinvestment, and exclusion. Geogr Compass 10(11):457–469
Werner M (2019) Geographies of production I: global production and uneven development. Prog Hum Geogr 43(5):948–958
World Bank (2020) World development report 2020: trading for development in the age of global value chains. http://documents.worldbank.org/curated/en/310211570690546749/pdf/World-Development-Report-2020-Trading-for-Development-in-the-Age-of-Global-Value-Chains.pdf
Yeung HW (2009) Regional development and the competitive dynamics of global production networks: an East Asian perspective. Reg Stud 43(3):325–351
Yeung HW (2016) Strategic coupling: east Asian industrial transformation in the new global economy. Cornell University Press

Sören Scholvin is an economic geographer from the Catholic University of the North in Antofagasta, Chile. He is also an associate at the 'Policy Research in International Services and Manufacturing' (PRISM) research unit at the University of Cape Town, South Africa. Sören's research deals with extractive industries and regional development in South America and Africa. He has published in peer-reviewed journals such as Area Development and Policy, Geografiska Annaler B and Resources Policy, among others. Sören is a co-editor of a special issues of Development Southern Africa (vol. 38, no. 1) and Growth and Change (vol. 52, no. 1).

Chapter 4
From Resource Peripheries to Emerging Markets: Reconfiguring Positionalities in Global Production Networks

Alexander Dodge

Abstract Several scholars have drawn upon the GVC/GPN approach to discuss the various dilemmas surrounding economic development in resource peripheries. However, most of the literature tends to focus on the "upstream" aspects of resource extraction and there have been few accounts surrounding market development in resource peripheries. In this chapter, I discuss how authorities in nation-states play a key role in reconfiguring the positionalities of resource peripheries from exporters to emerging markets within global production networks. I suggest that such positionalities are relationally constituted and therefore subject to change over time. These discussions are informed by an empirical study of natural gas-based energy development in Indonesia and Myanmar. Indonesia and Myanmar are both hydrocarbon-rich economies that have historically been coupled to natural gas production networks as exporters. The export of natural gas has occurred at the expense of domestic energy development as parts of the population have no access to affordable electricity supply. Recently, authorities in both countries have sought to reconfigure the current energy systems by increasing the utilization of natural gas for power generation by developing liquefied natural gas (LNG) infrastructure. This chapter finds there are significant challenges for LNG-based energy development in Myanmar and Indonesia which point to broader contradictions for resource peripheries surrounding the reconfiguration of their positionalities within GPNs.

Keywords Global production networks · Markets · State · Natural gas · Extractive industries

4.1 Introduction

A central tenet surrounding the notion of resource peripheries is that economic theories based upon research in economic cores are unable to grasp the peculiarities of economic development in resource peripheries. In their seminal paper, Hayter et al.

A. Dodge (✉)
Department of Geography, Norwegian University of Science and Technology, Trondheim, Norway
e-mail: alexander.dodge@ntnu.no

F. Irarrazaval and M. Arias-Loyola (eds.), *Resource Peripheries in the Global Economy*, Economic Geography, https://doi.org/10.1007/978-3-030-84606-0_4

(2003) propose a stakeholder theory that emphasizes the role of different stakeholder groups in remapping contemporary resource peripheries. Imbued in the literature surrounding research peripheries is a world-system theory tradition of distinguishing and relating core and peripheral nations/regions in the global economy. The notion of resource peripheries in Hayter et al. arguably finds its roots in the world-systems analysis of Wallerstein (1979) and others, even though they don't explicitly refer to this literature. Inherent within the core/resource periphery distinction is a central assumption that resources flow primarily from extraction sites in the periphery to markets in the core. In this sense, resource peripheries are conceptualized as being locked into structurally defined positions in core–periphery relations. Consequentially, the literature surrounding resource peripheries has yet to consider how the positionality of resource peripheries in the global economy comes to be reconfigured as resource peripheries that develop significant markets for the resources that they themselves produce.

In this chapter, I intend to contribute to the literature surrounding resource peripheries by discussing the peculiarities of market development in resource peripheries. Market development in core countries and the implications for global resources economies have been discussed in considerable detail (Bunker and Ciccantell 2005; Arboleda 2020; Urry 2013). However, thus far, the literature on resource peripheries has largely focused on resource peripheries as sites for resource extraction and has yet to inquire into resource peripheries as sites for resource consumption. Similar to Hayter et al. (2003) central argument that economic development in resource peripheries differs from that of cores, I suggest in this chapter that developing markets in resource peripheries are shaped by socio-spatial processes that differ from those that shape market development in cores. Specifically, I suggest that the socio-spatial processes shaping market development in resource peripheries should be framed within the context of how the positionality of resource peripheries in global extractive production networks comes to be configured and reconfigured over time.

The discussions in this chapter are further illuminated through case studies of market development for natural gas in Indonesia and Myanmar. Located in Southeast Asia, the countries of Indonesia and Myanmar have historically been considerable exporters of natural gas. These development trajectories have occurred at the expense of domestic energy development. I suggest that these development trajectories are a consequence of state strategies designed to consolidate power and enable capitalist accumulation within natural gas production networks. Furthermore, both state strategies and arrangements within natural gas production networks have been subject to change over time, which has led to a situation where authorities in Indonesia and Myanmar have sought to utilize liquefied natural gas (LNG) for energy development. This chapter finds that despite new state strategies and relationships within natural gas production networks, market development for natural gas is subject to different challenges that represent broader contradictions surrounding reconfiguring resource periphery positionalities in global production networks.

The rest of this chapter is structured as follows: In the next section, I discuss how market development in resource peripheries differs from that of cores and what it means to understand market development in the context of positionalities in global production networks. Methodology is then discussed. The next two sections detail how the positionality of Indonesia and Myanmar as resource peripheries in global natural gas production networks has been historically configured and how state strategies have been changing over time. The proceeding section then details how spatial and inter-organizational arrangements within global production networks are changing and opening opportunities for market development in emerging economies like Indonesia and Myanmar. The chapter then proceeds to discuss how Indonesia and Myanmar are reconfiguring their positionalities in natural gas production networks and the various contradictions related to these strategies. The chapter concludes by presenting an agenda for further research on market development in resource peripheries.

4.2 Resource Periphery Positionalities in Global Production Networks

Scholars focusing on market development for resources within world-systems theory have primarily focused on core countries. Bunker and Ciccantell (2005) examine how core economies (Portugal, the Netherlands, Great Britain, the United States, and Japan) have achieved and maintained global trade dominance by gaining control over the natural resources through cycles of accumulation. As their own national stockpiles of resources began to be depleted, core economies developed new technologies and infrastructure to exploit less accessible and more distant resources. The core economies secured access to cheaper, larger, and more stable supplies of resources by transforming the built environment both in their own countries and in extractive economies in ways that enabled economies of scale. Economies of scale were essential to reducing the input costs of resources in ways that expanded social production in these countries and enabled them to reach trade dominance.

Other scholars have highlighted the relationship between the ascendance of China as an economic power and its relationship to resource peripheries. Arboleda (2020) notes that Chinese investment strategies for primary production are markedly different from those of core economies such as Canada, Britain, and the United States. Arboleda suggests that China's rise "marks a new paradigm of nonhegemonic, multipolar, and "cooperative international relations" (p. 19). Others have noted how China formulated the *Zou Chuqu* (go out, go global policy) which instructed state-owned and private Chinese enterprises and banks to become closely involved in natural resource procurement and infrastructure development in resource peripheries to secure supply for mainland China (Mol 2011).

However, developing markets in resource peripheries can be different from core economies in several ways. First, whereas core economies rely on economies of scale to access distant resources, resources in resource peripheries are not distant, but in proximity. Second, economies of scale may not necessarily be the main imperative as demand centers in resource peripheries may be smaller and unable to offtake large volumes of natural resources. Third, the technology and capital for development often lie outside the control of the resource periphery, meaning that resource peripheries may be reliant on external actors to extract and produce resources (Bridge 2008; Arboleda 2020). Fourth, the export of resources generates significant rents and is a source of hard currency that can propagate existing regimes that allocate resources to exports rather than domestic development (Watts 2004; Beblawi and Luciani 2015). Finally, whereas core economies rely on technological/financial dominance and diplomatic capacity to secure access to resources, such means may nevertheless be limited in resource peripheries.

Despite the lack of technological, financial, and diplomatic capacities, resource peripheries do maintain sovereign control over the resources located within their territories (Bridge 2008). A key maneuver for resource-rich nation-states to develop markets for the resources located in their territories is to issue export restrictions, such as domestic market obligations. Several scholars have studied the role of export restrictions in the context of resource-based industrialization policy (Auty 1990; Neilson et al. 2020; Warburton 2017). Export restrictions provide preferential access to raw materials to domestic users, usually with the intent of developing local industry. For example, domestic market obligations for natural gas entail that the producers are obligated to sell a specified portion of their gas production to a rate equivalent to their export prices or at discounted rates to the domestic market (Saleh 2012). However, while nation-states may have the legal authority to reallocate domestically produced resources to domestic markets, their capacity to do so may be limited by the industrial organization and power relations in the global economy (Neilson et al. 2020). Understanding market development in resource peripheries must therefore go beyond notions of legal or territorial authority over natural resources and should instead be understood in the context of their positionality in the global economy, and how such positionalities are configured and reconfigured over time.

Thinking about resource peripheries in relation to their positionality entails accounting for how development trajectories within these places are shaped through the intersection of a variety of actors between places in the global economy. This notion of positionality draws upon the work of Sheppard (2002) who explains that positionality entails mapping the shifting, asymmetric, and path-dependent relationships between different agents and how these relationships work to articulate places within broader structures in the global economy. Working through the notion of positionality as a relational attribute allows for a more detailed examination into the socio-spatial processes that produce different displacements and marginalization in resource peripheries, and how these processes may change over time. For example, Kortelainen and Rannikko (2015) have drawn upon Sheppard's ideas surrounding positionality to consider how abrupt shifts in the direction of relations surrounding resource peripheries alter their positionalities. In their research on Russian forestry communities located in the Finnish-Russian borderland, Korelainen and Rannikko demonstrate how the positionality of these communities shifted dramatically after sudden shifts in border regulation, transportation connections, ownership arrangements, and political organizations.

The positionality of resource-peripheries can thus be considered a relational attribute that is constituted through uneven and evolving power relationships between places within the global economy. While governments in resource peripheries may have the legal authority to issue export restrictions on the resources embedded within their territories, these resources are still embedded within infrastructural, financial, and technological networks that extend both within and beyond these territories (Arboleda 2020; Allen 2016). These networks can both enable and restrict the capacity of governments to allocate such resources to domestic development.

For the sake of developing more analytical precision in thinking through the positionality of resource peripheries, this chapter suggests drawing upon the global production network (GPN) framework. Utilizing the GPN framework is well in line with the work of several scholars focusing on production networks in extractive industries (Bridge 2008; Barratt and Ellem 2019; Dodge 2020).[1] For example, Bridge (2008) suggests that the GPN approach can be used as a heuristic tool to understand the complex linkages between the organization of production in the extractive industry and its implications for the development opportunities available to the territories coupled to these networks. The GPN framework sheds light on how significant relationships between actors on different scales shape how territories are embedded in networks over time.

In this chapter, I focus specifically on the relationships between nation-states and firms within GPNs, and how these relationships shape the positionality of resource peripheries. My focus on firm–state relationships is well in line with several scholars who have called for greater attention to the state within the GPN framework (Glassman 2011; Smith 2015; Yeung 2016; Dodge 2020). Smith suggests that greater theorization of the state in GPNs would help researchers better account for the institutional factors that shape the governance of global production networks. Drawing upon Jessop's (2007) strategic-relational approach to state theory, Smith (2015) conceptualizes state strategies in GPNs as articulated through particular modes of development and accumulation regimes intended to consolidate power and stabilize social struggles within the state.

Thinking about the positionality of resource peripheries in global production networks, and drawing upon a strategic-relational approach to state theory, opens for thinking through the temporality of market development in resource peripheries. Barratt and Ellem (2019) suggest that transformations in the arrangement of global production networks, as well as inter-relationships between firms, states, and labor,

[1] For the sake of the arguments presented in this chapter, I adopt the heuristic approach of GPN 1.0 framework rather than the more theoretical approach of GPN 2.0 developed by Coe and Yeung (2015). As a heuristic device, rather than a theory, the GPN 1.0 framework provides a useful set of concepts which can be used to describe the empirical reality of global production networks, including identifying a key set of power relationships between different actors and how the territorial, societal, and networked embeddedness of these relationships changes over time (Hess 2004). Rather than attempting to identify the causal mechanisms through which power relationships in GPNs emerge, my focus here is on detailing the empirical reality of these power relationships and how different entities are situated within these power relationships in space/time. This approach falls in line with the notion of positionality detailed by Shepperd (2002).

should be understood as spatio-temporal processes that evolve over time. The positionality of resource peripheries in GPNs is likely subject to change as both the industrial organization of GPNs and state strategies are transformed over time. This has been the case in Myanmar and Indonesia, where transformations in liquefied natural gas production networks and shifts in state strategies have had implications for market development in these countries.

4.3 Methods

The empirical material for this chapter is based upon an analysis of different institutional and consultancy reports as well as academic research on the political economy of resources in Indonesia and Myanmar (Seah 2014; Dodge 2017; Simpson 2016; Warburton 2017; Choy 2011; Dobermann 2016). In addition, during fieldwork in Indonesia, Myanmar, and Singapore in 2016 and 2017, the author interviewed (semi-structured format) 18 representatives from LNG-related businesses, law firms, and consultancies familiar with the situation surrounding natural gas market development in Indonesia and Myanmar. Three of these representatives were interviewed in Indonesia, three in Myanmar, and 12 in Singapore. Most of the interviews were conducted in Singapore as most LNG-related businesses in Southeast Asia have their regional headquarters in Singapore (Fig. 4.1).

4.4 Indonesia: Becoming the World's Largest Liquified Natural Gas Exporter

The exclusion of resource peripheries from markets for the resources that they themselves produce is a consequence of how such resources are made available to global markets. In his comparative study on the natural gas industry in Boliva and Peru, Irarrázaval (2020) explains that the articulation of resource-holding regions in natural gas production networks is underpinned by a complex and contingent articulation of socio-ecological relations that makes natural gas into a resource. Irarrázaval notes that states play a crucial role in attracting and facilitating resource extraction as they deliver infrastructure, allocate exploitation rights, and provide the conditions for the large capital investments that make the commoditization of natural gas resources possible. Similarly, as will be detailed below, the Indonesian government played a crucial role in making natural gas a resource that could be exported. However, the export of Indonesian natural gas came at the expense of domestic energy development.

From the 1960s to 1970s, natural gas played a minor role in the Indonesian oil industry. Although some gas was used for fertilizer plants, the majority was discarded through flaring or venting (Bee 1982). Foreign direct investment in the Indonesian oil

Fig. 4.1 Map of Southeast Asia showing the location of Myanmar and Indonesia

and gas industry had been significantly reduced after nationalization in 1957 (Lindblad 2015). This changed in 1967 when President Suharto provided fiscal incentives for foreign direct investments, which resulted in Indonesia becoming a key oil and gas producer in Southeast Asia. In 1971, when the Mobil Oil Corporation discovered the Arun natural gas field in Northern Sumatra (Mehden and Lewis 2006), at the time, natural gas around the world was mainly transported through pipelines; however, President Suharto's regime perceived that the Arun gas field was too far from large population centers and the demand was too low to warrant the development and financing of pipelines. Instead, natural gas from the Arun field was cryogenically liquefied and exported to high-income consumers in Japan. Mehden and Lewis (2006) explain that the Arun gas field has proven to be one of the most lucrative LNG operations in the twentieth century. The simultaneous development of the Bontang LNG terminal would lead Indonesia to become the largest exporter of LNG in the world, a position that the country held until 2006 (Seah 2014). However, the coupling of Indonesian natural gas resources to export-oriented production networks occurred at the expense of domestic energy development (Mehden and Lewis 2006).

State strategies related to coupling to export-oriented production networks at the expense of domestic energy development can be related to what Jessop (2007) terms "state strategic selectivity". State strategic selectivity refers to an examination of institutional forms within the state that end up privileging some actors, strategies, actions, etc. over others with the intention of stabilizing social struggles and consolidating political power. By coupling nationalized natural gas assets to LNG production networks, Suharto's regime consolidated political power through a statist mode of development as revenues were directed to rural development schemes, infrastructure spending, and fuel oil subsidies (Aspinall 2013). Suharto's power was based on a system of patronage under the New Order Regime which involved capturing local elite support, such as awarding infrastructure projects to crony businessmen (Mackie 2010).

There are two reasons why coupling to export-oriented LNG production networks was privileged over domestic market development. The first key reason is related to the complicated physical properties of natural gas and the nature of relations in primary resource sectors. Bunker and Ciccantell (2005) explain how the globalization of resources creates a trend toward economies of scale. This was the case for natural gas, where the complicated physical properties of natural gas require large and lumpy capital investments. LNG facilities are largely capital-intensive, and to reduce the per-unit costs of LNG and generate greater profits, LNG producers relied on economies of scale. Consequentially, LNG markets were primarily exclusive to core economies with large urban demand centers such as Japan, and lower demand centers in Indonesia were excluded.

The second key reason why export markets were privileged was that Indonesian authorities sought to limit their financial involvement in LNG production facilities while at the same time the Japanese government actively supported domestic market development in order to reduce the use of expensive fuel oil in the electricity generation mix (Mehden and Lewis 2006). For example, the Japanese export–import bank loaned funds for Indonesian LNG projects. The basis for investments in these large, capital-intensive projects was through take-or-pay obligations in gas supply agreements (Corbeau and Ledesma 2016). By agreeing to take-or-pay obligations, customers, usually regulated gas monopolies, would commit to offtaking volumes for a period of time, usually 15–20 years, or pay a significant fine. The consequence of these arrangements is that they relied on the creditworthiness of consumers to agree upon such contract, and often domestic Indonesian buyers were considered insufficiently creditworthy to provide such guarantees (Corbeau 2016). Therefore, domestic markets in Indonesia were historically excluded from the LNG production networks.

While the state plays a key role in establishing the conditions by which export-oriented natural gas production networks emerge, these relations are subject to change. After the Asian economic crisis of 1997, the economy fell back to commodity exports and Suharto's authoritarian regime fell. At the same time, infrastructure spending dropped and remained stagnant and the country became more dependent upon resource exports (Negara 2016). This produced a significant amount of discontent in Indonesian society as large regional disparities persisted between central parts of Java and the islands in the east. These regional disparities not only include disparities in terms of regional GDP per capita but also in terms of infrastructure development, electricity usage, and poverty rates—among other aspects (Nugraha and Prayitno 2020). According to Jessop (2007) the strategic selectivities of the state that reflect and modify the balance of class forces are subject to change over time, as new social struggles threaten current hegemonies. In the case of Indonesia, discontent with growing regional disparities led to new political coalitions with the election of Joko Widodo "Jokowi" (Warburton 2016).

In 2015, Jokowi was elected upon a campaign with the aim of reducing poverty in the rural parts of the country. Warburton (2016) characterizes Jokowi's approach to governance as appeasing a nationalist, state-centric development narrative, focused on reducing unprocessed mineral exports and enhancing the countries industrial capacity through infrastructure development. During his 2015 presidential run, Jokowi promised to relieve poverty in Indonesia by "modernizing" physical infrastructure, particularly in rural Indonesia (Yusuf and Sumner 2015). A key contentious issue at the time was the significant dependence upon fuel oil (nearly 83%) in the electricity generation mix in the more rural islands in the eastern and central parts of the country (ESDM 2016). Fuel oil in Indonesia is heavily subsidized and the high costs of fuel oil are a hinder to increasing the electricity generation capacity needed for industrial development (Seah 2014).

In 2015, Jokowi's regime launched a 35 Gigawatt (GW) fast-track electrification program to be installed by 2019, 13 GW of which was to be supplied by natural gas (see Fig. 4.2). In doing so, Jokowi planned to utilize a larger portion of the country's domestic natural gas production. These strategies can be considered within the context of new strategic selectivities in the state, where accumulation through industrial development becomes privileged over traditional interests surrounding the

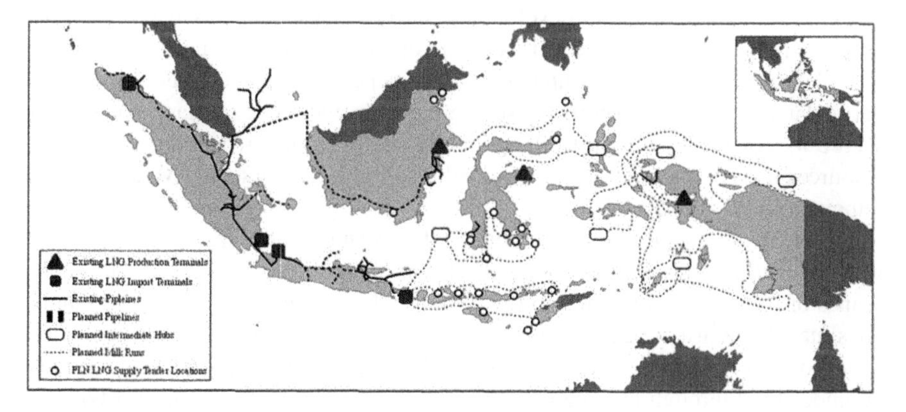

Fig. 4.2 Map of Indonesia showing current and planned natural gas pipeline infrastructure under the 35 GW fast track electrification program

export of resources such as natural gas. However, such changes cannot be understood in the light of state strategies alone, and they must be considered within the context of emerging arrangements within global production networks. In the following sections, I will discuss how changes in spatial and inter-organizational arrangements in LNG production networks opens opportunities for Indonesia to reconfigure its positionality in LNG production networks. But first the relationships through which the positionality of resource peripheries is configured will be further discussed through the study of natural gas exports in Myanmar.

4.5 Myanmar: Consolidating the Power of Military Rule

The positionality of a country within global production networks is not only constituted through the relationships between transnational corporations and nation-states but also through geopolitical relationships between different states. According to Glassman (2011), a study of geopolitics presents a set of agents and agencies that are not merely exterior to GPNs, but play a central role in the constitution of such networks. Myanmar is centrally located between three large market economies—China in the north, Thailand in the east, and India to the west. At the same time, Myanmar is the 16th largest natural gas exporter in the world (2017 est.) with significant potential for further exploration and development (CIA 2020). Consequentially, Myanmar is enmeshed within a network of geopolitical relationships between its neighboring countries due to its strategic location and resources (Kolås 2007).

While Myanmar's strategic location and natural gas reserves could serve as the foundation for economic growth and improvement in living standards for the country's population, Myanmar remains the poorest economy in Southeast Asia (Stokke et al. 2018). Myanmar's economy has been heavily sanctioned and plagued by violence and internal conflicts at the hands of the ruling military government (Stokke et al. 2018). Resource development surrounding natural gas in Myanmar has been marred by violence and controversy. In order to maintain its power and regime of violence, the military government has been dependent upon commodity exports, particularly natural gas exports, to secure hard currency. The military's attempt to secure its power in addition to the geopolitical situation surrounding oil and gas resources in Myanmar has largely entailed that domestic markets in Myanmar have remained largely undeveloped.

Watts (2004) suggests that the ability of oil (and natural gas by extension) to generate conflicts should be understood in terms of the already existing political landscapes of forces, identities, and forms of power. Natural gas resources in Myanmar are situated within a pre-existing violent oppressive regime, and the socio-ecological conditions through which natural gas in Myanmar is made into a resource must be understood through this regime. Namely, natural gas development in Myanmar must be understood in the context of the militarization of Myanmar society and the prevalence of conflicts in Myanmar's contested borderland areas.

4.5.1 Contested Borderlands

Myanmar is a multi-ethnic society where the Bamar majority is largely concentrated in the center of the country and the non-Bamar minorities live primarily along the mainland borders of the country (Lertchavalitsakul and Meehan 2020). Following independence in 1948, post-colonial governments unsuccessfully attempted to consolidate the power of the state by asserting control through political negotiation over the country's borderlands. After the 1962 military coup, the military junta justified the coup through claims that the country's constitutional democracy was incapable of protecting the country from external threats and to "control the unruly borderlands and threat posed by disloyal ethnical minority population" (Lertchavalitsakul and Meehan 2020, 206).

While the military junta sought to establish greater authority throughout the country through violence against ethnic minority populations, the mountain ranges and dense forests that encompass much of the countries border regions, coupled with limited road networks, limited military activities in these areas. Some of these contested areas are controlled by different ethnic armed organizations, such as the Kachin Independence Army along the northern border with China and Karen National Liberation Army along the eastern border with Thailand. A major source of income for the Ethnic Armed Organizations (EAO) was from cross-border trade in prohibited goods and smuggling as well as collecting tax from mining firms in areas that were not under full government control.

The borderlands became increasingly militarized as the military sought to assert control over natural resources and cross-border trade. Although the military government brokered ceasefire agreements with some EAOs such as those in the Shan State and Kachin State in the late 1980s, it continued its attacks on non-ceasefire groups—particularly against the Karen National Union in the Kayin State. Ethnic borderland populations were subject to military abuses and violence such as forced portering for the army, sexual violence, forced purchase of goods below the market price, and arbitrary systems of local "taxation". Human rights abuses by the military led the western governments to impose sanctions against Myanmar. Despite sanctions, the military government would remain in power largely through rents generated through natural gas exports to neighboring countries.

4.5.2 Development of the Yadana Pipeline

In 1997 and early 1998, Myanmar was in a financial crisis due to international sanctions, economic mismanagement, and the Asian financial crisis. The value of the Burmese Kyat plummeted in 1997 and the countries foreign exchange reserves shrunk to less than the foreign currency deposits they are supposed to cover (EarthRights International 2008; The Irrawaddy 1997). In desperate need of hard currency to continue to finance the military junta, the government turned to natural gas.

In the 1980s, the Myanmar government invited foreign companies to bid for offshore exploration in 18 concession blocks in the Gulf of Martaban (Kolås 2007). Two major offshore gas fields, Yadana and Yetagun were discovered in the Gulf of Martaban. In 1992, the Myanmar Oil and Gas Enterprise and the major oil company, Total, signed a memorandum of understanding regarding the exploitation of the Yadana offshore gas field (Holliday 2005). The Petroleum Authority of Thailand (PTT) contracted to purchase at least 80% of annual output and joined the project as a major stakeholder. The politics around the Yadana Gas Field was highly contentious. The ruling military regime was notorious for committing multiple human rights violations and had been strongly condemned by the United Nations Human Rights Commission and U.S. State Department (Larsen 1998).

State authority in Myanmar was expressed through violence, and pipeline development has been marred by controversy and human rights abuses. The Yadana pipeline would run across the Tenasserim division, a territory that the Burmese military had been fighting with Armed Karen and Mon EAOs. The eastern sector of the pipeline's path was technically designated a war zone by the SLORC. Karen forces threatened to sabotage the construction of the pipeline and several insurgent attacks had already taken place. To secure the area around the pipeline the military used forced labor to transport supplies for troops for military offensives, building security outposts, building railways, among other activities (Larsen 1998). In addition, the military used torture, rape, and unlawful land seizures to remove villagers from areas slated for development, and also the military's use of unpaid, forced labor to facilitate the pipeline construction (Simpson 2016).

Revenues from natural gas export to Thailand, and subsequent exports to China after the development of the Shwe gas field, were a major source of foreign currency for the military junta. In 2010–2011, gas accounted for approximately 3 billion gross revenue (Simpson 2016). These revenues supported the expansion of the military and its ability to harass its opponents, while little was spent on health, education, and energy development. The military went on an accelerated arms-buying spree, upgrading navy and air force weapons and increasing the size of its army. Instead of using natural gas to build up the domestic industry, natural gas in Myanmar has been captured by a regime of exportism designed to perpetuate military rule (Larsen 1998; Carroll and Sovacool 2010; Simpson 2016). Furthermore, natural gas helped insert Myanmar into a web of geopolitical relationships with its neighboring countries who provided economic connections and diplomatic support in the face of international sanctions (Kolås 2007). Consequentially, domestic market development for natural gas was systematically underprioritized, with the exception of limited supply to Yangon, and up until 2011 the majority of the population has had no access to electricity (World Bank 2018).

However, such strategic selectivities are subject to change. In 2008, military rulers made a new constitution that opened up for political liberalization while at the same time institutionalizing power for the military (Stokke et al. 2018). The 2008 constitution provided a basis for elections to local and parliamentary elections in 2010 and 2015. While the 2010 elections were flawed, they were used to transfer power to a nominally civilian government led by President Thein Sien and the military's

Union Solidarity and Development party (USDP). The USDP government initiated a series of norms in favor of formal democracy, open economy, and ceasefire agreements with EAOs most with the intention of normalizing diplomatic and economic relations and attracting foreign direct investments. The reforms eventually led to the first free general election in 2015.

4.5.3 Election of NLD

In November 2015, a free general election was held which resulted in a landslide victory for the National League of Democracy (NLD) and a defeat for the USDP. NLD ran on a platform of finding durable political solutions to long-standing tensions between the central state and the country's ethnically diverse borderlands that would provide the foundations for peace and reconciliation. However, such efforts have largely stalled due to the violence mobilized by the army against the Rohingya population in Rakhine state in 2016/7 and intensified counterinsurgencies in the Kachin and North Stan State.

Since opening up in 2012, the economy of Myanmar has grown considerably which has resulted in a significant increase in energy demand. Consequentially, the NLD has faced the prospect of impending power shortages, particularly in the city of Yangon. The primary energy source for electricity in Myanmar is hydropower, the capacity of which is significantly reduced during the dry season, often resulting in energy shortages and blackouts. The city of Yangon is also connected to natural gas pipelines from natural gas fields; however, the current domestic gas supply is not enough to cover the demand of power stations. Although natural gas production in Myanmar is more than enough to meet energy demand needs, most production is reserved for long-term gas supply agreements with Thailand and China and only one-fifth is available for domestic markets.

A USTDA study in 2017 reported that unserved power shortages of over 20% are likely from 2020 to 2022 unless the government is able to develop power stations in the near future (Kean 2017). The NLD is under immense pressure to solve impending energy shortages, as it would threaten the NLD's electoral base in 2020. Furthermore, the Myanmar military still holds a mandated 25% of the seats in parliament and is working to undermine the capacity and legitimacy of the NLD government (Stokke et al. 2018). These political dynamics in Myanmar, therefore, point to new strategic selectivities, where the government is likely to prioritize energy development to sustain economic growth and maintain political power. Thus, Myanmar finds itself in a similar situation as Indonesia, where developing energy markets becomes a primary concern for the state. As, will be discussed in the following sections, these political drivers in combination with changes in LNG production networks lead to a situation where both Myanmar and Indonesia are becoming emerging markets for liquefied natural gas.

4.6 Evolving Liquefied Natural Gas Production Networks

As mentioned previously in the discussion on LNG exports in Indonesia, LNG production networks have historically been exclusive to high-income markets in North Asia. Such spatial outcomes have been a consequence of the project-like character of production networks, so the focus is on economies of scale to achieve cost reduction and the use of take-or-pay contracts. However, while these outcomes may have historically excluded market development in developing economies like Myanmar and Indonesia, LNG production networks are currently evolving in ways that are diversifying markets toward lower-income economies.

By understanding LNG production networks in the context of time, we can better understand the implications of evolving spatial and organizational arrangements for the positionality of Myanmar and Indonesia within these GPNs. In their analysis of LNG production networks, Bridge and Bradshaw (2017) note that LNG production networks are evolving, from what they term as a:

Floating pipeline model of point-to-point, binational flows orchestrated by producing and consuming companies and governed by long-term contracts, to a more geographic and organizationally complex production network that is constitutive of an emergent global gas market (1).

Bridge and Bradshaw's analysis of LNG production networks reflects a situation where LNG production networks are becoming more spatially diversified and organizationally fragmented and, as a result, is enabling the capacity of production networks to achieve greater flexibility and reduce the need for take-or-pay commitments. The dynamics are the result of a significant wave of investments in LNG production capacity in the USA, Australia, and Qatar that were made without securing long-term take-or-pay obligations while at the same time, LNG demand growth in mature importing countries like Japan, Korea, and China has been slowing. The consequence of these developments is that LNG producers have looked to emerging economies as potential markets for LNG surpluses.

Typically, importing LNG gas is a cost-intensive venture, as LNG must be stored and regasified at land-based terminals before it can be sent to power plants for electricity generation. However, several companies have designed and developed ways to reconvert existing LNG carriers into floating storage and regasification units (FSRU). Unlike land-based terminals which must be developed onsite and often required skilled contractors which can be expensive to send to more rural demand centers in emerging economies, FSRU's can be developed in pre-existing shipyards in Singapore or South Korea. This significantly lowers the threshold and price of importing LNG in emerging economies. The FSRU technology in combination with more flexible contracting arrangements opens the opportunity for emerging economies to import LNG. As I will demonstrate in the following sections, these changes have opened the opportunity for Indonesia and Myanmar to reconfigure their positionalities within natural gas production networks.

4.7 New Development Trajectories

Whereas resource peripheries in Indonesia and Myanmar were previously excluded from LNG production networks, evolving spatial and organizational arrangements in combination with shifting strategic selectivities have led authorities in these countries to reconfigure their positionality in LNG production networks.

4.7.1 Indonesia

Jokowi's plans to significantly expand electrification through the development of natural gas infrastructure comes at a time when domestic production has been steadily on the decline since peak production in 2010 (Seah 2014). Although Indonesia has historically been a net exporter of natural gas, domestic demand has continuously exceeded export volumes since 2012 (Seah 2014). Furthermore, export volumes have declined 25% from their peak in 2010 until 2019. In combination with the fall of natural gas prices since 2014, revenue from the industry has steadily declined as a portion of government income. According to the Indonesian government's own estimates, Indonesia will become a net importer by 2028 and will likely cease all gas exports by 2034 (Gomes 2020). Thus, Indonesia is in the midst of a positionality switch in natural production networks, where it may become a significant importer of LNG rather than an exporter as it has been in the past.

The ongoing positionality switch is a consequence of government policy that predates Jokowi's regime, which has sought to allocate domestic natural gas production to domestic markets. In 2004, the Indonesian government established obligations (GR 35/2004) for upstream entities to allocate 25% of gas production for domestic consumption (Gomes 2020). Since then, domestic natural gas demand growth has increased due to industrialization and urbanization, particularly in western Java; however, at the time, natural gas was supplied through domestic pipelines. Gas shortages in fields supplying Java led the government to consider building LNG terminals to utilize domestically produced gas.

In 2008, the government commissioned the countries first import terminal to be in operation by 2012 (Purwanto et al. 2016). The project took an existing LNG carrier that was built in 1977 and converted it into an FSRU that would be used to supply west Java. Since 2012, three more LNG import terminals have been built. At the same time, several LNG production terminals have started to experience a decline in production, due to aging gas fields. The depletion of gas fields connected to the Arun LNG production plant caused the plant to cease operations in 2014. The plant was subsequently converted into an LNG receiving terminal that would supply neighboring industries.

Most of the countries' LNG receiving terminals have been located around Java, while infrastructure in the rest of the country remained undeveloped. As part of Jokowi's 35 GW fast-track electrification plan, the Indonesian government launched plans to develop natural gas infrastructure in the eastern and central parts of Indonesia to reduce the use of expensive fuel oil and increase energy access in these areas. A barrier toward developing natural gas infrastructure in Indonesia is that the population is fragmented and spread across numerous islands and the total potential demand of consumption centers on these islands is too low to generate adequate rates of return for LNG suppliers. Consulting companies, such as DNV-GL, have suggested that the government could aggregate the demand of multiple islands by establishing small LNG milk-runs (Choy 2011). Milk-runs entail developing intermediate storage hubs for the delivery of large LNG cargos and then aggregating the demand of smaller, more remote terminals through small, point-to-point, LNG logistic chains. Despite the feasibility of developing LNG markets in Asia, plans to develop small LNG terminals in Indonesia have been repeatedly delayed.

Development of LNG receiving terminals at main demand centers, domestic obligations, and plans to develop small LNG terminals in the eastern and western parts of Indonesia represent an ongoing positionality switch within global natural gas production networks. This positionality switch occurs through new relationships within the production network itself. However, despite these emergent relationships, there are two key challenges related to plans to reconfigure Indonesia's positionality in LNG production networks by significantly investing in LNG infrastructure throughout the country.

The first key challenge is that while supplying LNG to the smaller demand centers in the central and eastern parts of Indonesia may be technologically and economically possible, there are numerous supply and economic risks involved in doing so. According to an interview with a representative from an LNG-related company, establishing intermediate hubs and terminals in multiple locations increases the total amount of LNG stored in the system, thereby expanding storage costs. In addition, offloading a single carrier at multiple ports increase the daily shipping costs. Thus, for milk-runs to work, shipping and storage need to be optimized and tightly coordinated. However, the systematic coordination of LNG chains contradicts the increasing organizational fragmentation and flexibility in contracting arrangements between partners involved in LNG projects. Therefore, for small LNG projects to be possible, producers required long-term commitments and guarantees from the government. However, this poses a considerable risk to government partners, particularly if natural gas prices rise or if natural gas demand is significantly reduced. Thus, while using LNG for energy development in Indonesia is possible, it is not necessarily suitable for energy development needs in the region.

The second key challenge is that tariffs, which are set by the government, for natural gas in Indonesia are significantly lower than prices within international markets (Seah 2014). In other words, much greater profits can be earned by producers through exporting natural gas, than for selling natural gas in domestic markets. The government could increase tariff prices, but this would in turn make energy more unaffordable for households and industry (Gomes 2020). Given that the government is looking to reduce spending on subsidies by transitioning away from fuel oil, subsidizing natural gas prices would contradict these aims (Seah 2014). Given low tariffs, the consequence of government policies to increase domestic natural gas production is that it reduces the attractiveness of investment in exploration and development, which in combination with maturing fields leads to a situation where the country may end up importing natural gas from abroad (Gomes 2020).

Thus, the Indonesian government stands in a dilemma, whereby increasing the domestic utilization of domestically produced natural gas, reduces investments in gas fields, which in turn leads to a situation where Indonesia could become a net importer of natural gas. The consequence then is that the country could end up relying on expensive LNG imports based on a policy intended to utilize the countries domestic resources for industrial development.

4.7.2 Myanmar

Although domestic natural gas production in Myanmar is more than enough to meet Myanmar's immediate energy needs, most of this gas production is locked into long-term gas supply agreements with Thailand and China (Dodge 2017). Much of Myanmar's deepwater areas remain largely unexplored and could prove to hold large natural gas deposits; however, there has been little investment due to previous sanctions. Although sanctions have been lifted following the 2015 general elections, Myanmar has struggled to attract investment due to domestic market obligations, low global oil and gas prices, and disagreements surrounding production sharing agreements (Dodge 2017). Furthermore, even if new offshore discoveries were made, they would be unlikely to meet Myanmar's impending power shortages, given the time it would take from the first discovery to production. Consequently, the government has turned toward plans to import LNG while continuing to export domestically produced natural gas to China and Thailand.

Plans to import LNG were originally sponsored by the World Bank. In 2013, the World Bank initiated the Myanmar Electric Power Project for Myanmar with a total project cost of 140$ million, with the aim to increase the capacity and efficiency of gas-fired power generation. In 2016, the World Bank commissioned consultants to examine potential sites for a large-scale floating storage and regasification project (FSRU) to supply natural gas to the main city of Yangon (Mainhardt 2019). However, there were numerous challenges regarding the development of infrastructure, as the shores of Yangon were too shallow for large LNG carriers to enter and extensive pipeline development would be needed to supply natural gas to power plants in Yangon (Dodge 2017).

According to reporting in *Frontier Myanmar*, the World Bank eventually advised the government against an FSRU and advocated for a small LNG solution where small ships would deliver LNG directly to a site in Yangon (Kean 2018). However, contradictory to the advice of the World Bank, the Myanmar government in 2018 made a surprise announcement that it had agreed to notices to proceed for two large and one small FSRU projects. Furthermore, whereas the World Bank had envisioned a competitive tender process for LNG projects, the government proceeded instead with bilateral agreements with different parties on its own initiative, without seeking the advice of the World Bank.

A key challenge for LNG projects in Myanmar is that current tariffs in Myanmar are too low to reflect the costs of LNG imports and related infrastructure, and therefore Myanmar would likely need to raise the tariffs for LNG projects to be possible. A report by Bank Information Center Europe (Mainhardt 2019) criticized the plan to importing LNG because it would make electricity prices for ordinary households in Myanmar largely unaffordable, and it is unclear if the government could afford targeted subsidies to these households. Furthermore, according to interviews with lawyers in Myanmar consulting LNG companies, suppliers were likely to require long-term guarantees for LNG projects, despite increasing flexibility in LNG production networks, due to the need for additional investments in associated power plants and infrastructure. Long-term agreements would likely lock in Myanmar's energy system into expensive LNG imports, even if domestic gas becomes available in the future. Consequentially, Myanmar would not be able to offtake domestic production if new fields are developed in the future, and future production would likely be exported rather than utilized domestically.

4.8 Conclusion

In this chapter, I have discussed the need to go beyond structural perspectives on the core–resource periphery divide and has instead suggested a relational understanding of the positionality of resource peripheries in global production networks. In doing so, I have aimed to demonstrate how resource peripheries come to reconfigure their positionality within GPNs from resource exporters to emerging markets. I have primarily focused on how the relationships between the nation-state and firms in GPNs shape these positionalities. These theoretical discussions have been illuminated through case studies on plans for liquefied natural gas (LNG) market development in Indonesia and Myanmar—two countries that have historically been significant exporters of natural gas and are now looking toward LNG as a means toward energy development. The case studies are compared in Table 4.1.

The first part of the empirical section in this chapter, I pointed to the strategic selectivities of the nation-state and the state's relationship to firms in natural gas production networks as a means of accounting for why Indonesia and Myanmar have exported natural gas at the expense of domestic energy development. These strategic selectivities can be labeled as *exportism.* In Indonesia, revenues from LNG exports have been a means of securing rents that could be then utilized by the Suharto regime to consolidate political power through a patronage system designed to capture local elite support. In combination with the organization of LNG production networks around economies of scale and long-term gas supply agreements with creditworthy buyers in North Asia, these dynamics resulted in the exclusion of small demand centers in central and eastern Indonesia from natural gas markets. In Myanmar, the ruling military regime depended upon natural gas exports as a means of securing

Table 4.1 Comparison of state strategies and contradictions surrounding natural gas market development in Indonesia and Myanmar

	Indonesia	Myanmar
Strategic selectivities		
Exportism	Rents generated from LNG exports was used to finance system of patronage under Suharto regime	Rents from pipeline exports was used to replenish foreign currency reserves and finance military spending
Developmentalism	Develop markets for LNG in peripheral regions to reduce reliance on imported fuel oil	LNG imports is considered as a means to resolve immediate energy needs
Contradictions surrounding the import of LNG		
	Numerous supply and economic risks involved in supplying LNG to smaller demand centers	Would make electricity prices for ordinary households unaffordable, requiring targeted subsidies
	Increasing utilization of domestically produced natural gas reduces investments in gas fields, resulting in a situation where the country needs to rely on LNG imports	Would lock energy system into long-term LNG imports, even if domestic gas becomes available in the future

revenue and maintaining geopolitical ties with its neighboring countries to secure its power and regime of violence at the expense of domestic energy development. However, while these strategic selectivities resulted in the exclusion of domestic markets from natural gas, these selectivities have been subject to change over time.

Both Myanmar and Indonesia have been subject to highly turbulent political changes within the last decade, which have led these countries to pursue domestic energy development as an avenue for maintaining political power. Shifting strategic selectivities in combination with changes in organizational and spatial arrangements in LNG production networks have led authorities in both countries to consider reconfiguring their positionalities in LNG production networks from primary exporters, to emerging markets. These strategic selectivities can be labeled as *developmentalism.* To develop markets for natural gas, governments in Indonesia and Myanmar have relied on a new set of alliances within LNG production networks—particularly with actors who can secure investment in and procure floating storage and regasification units (FSRUs)—which are more affordable than land-based storage and regasification terminals. The drive to develop markets in countries like Indonesia and Myanmar is a result of new market development imperatives in LNG production networks following a surplus of LNG production capacity on global markets.

However, such attempts to shift positionalities within natural gas production networks are not without contradictions. A key contradiction is that both Myanmar and Indonesia may end up importing natural gas from other countries, while simultaneously exporting domestically produced natural gas. In Indonesia, the government has attempted to secure domestic natural gas supply by issuing domestic market obligations to producers. However, low tariffs designed to make energy affordable to low-income households make supplying domestic markets much less profitable than exporting natural gas to high-income markets. Consequentially, domestic market obligations reduce incentives to attract investment in the development of new natural gas fields, which can result in a situation where Indonesia must import LNG to meet domestic market demand. In Myanmar, domestically produced natural gas is locked into long-supply agreements with China and Thailand, and therefore the country is looking toward importing LNG. However, developing LNG infrastructure would likely require long-term commitments which could lock the countries energy system to imported LNG, even as new domestic gas resources become available.

Contradictions surrounding reconfiguring positionalities in LNG production networks point to the peculiarities of developing markets for resources that resource peripheries themselves produce. The contradictions become apparent when accounting for how the positionalities of resource peripheries in global production networks are relationally constituted and reconfigured over time.

References

Allen J (2016) Topologies of power: beyond territory and networks. Routledge, New York

Arboleda M (2020) Planetary mine: territories of extraction under late capitalism. Verso Trade, London

Aspinall E (2013) A nation in fragments. Critical Asian Stud 45(1):27–54. https://doi.org/10.1080/14672715.2013.758820

Auty RM (1990) Resource-based industrialization: sowing the oil in eight developing countries. Oxford Unversity Press, Oxford

Barratt T, Ellem B (2019) Temporality and the evolution of GPNs: remaking BHP's Pilbara iron ore network. Reg Stud 53(11):1555–1564. https://doi.org/10.1080/00343404.2019.1590542

Beblawi H, Luciani G (2015) The rentier state. Routledge, New York

Bee OJ (1982) The petroleum resources of Indonesia. Springer Netherlands, Amsterdam

Bridge G (2008) Global production networks and the extractive sector: governing resource-based development. J Econ Geogr 8(3):389–419

Bridge G, Bradshaw M (2017) Making a global gas market: territoriality and production networks in liquefied natural gas. Econ Geogr 93(3):215–240

Bunker SG, Ciccantell PS (2005) Globalization and the race for resources. JHU Press, Baltimore

Carroll T, Sovacool B (2010) Pipelines, crisis and capital: understanding the contested regionalism of Southeast Asia. Pac Rev 23(5):625–647

Central Intelligence Agency (CIA) (2020) Natural gas exports. In: The world factbook

Choy V (2011) Oppurtunities and risks of small scale LNG development in Indonesia. DNV-GL, Singapore

Coe N, Yeung H (2015) Global production networks: theorizing economic development in an interconnected world. Oxford University Press, Oxford

Corbeau A (2016) LNG contracts and flexibility. In: Corbeau A, Ledesma D (eds) LNG Markets in transition: the great reconfiguration. Oxford University Press, Oxford, pp 502–553

Corbeau A, Ledesma D (2016) LNG markets in transition: the great reconfiguration, LNG markets in transition: the great reconfiguration. Oxford University Press, Oxford

Dobermann T (2016) Energy in Myanmar. In: International growth centre

Dodge A (2020) The Singaporean natural gas hub: reassembling global production networks and markets in Asia. J Econ Geogr. https://doi.org/10.1093/jeg/lbaa011
Dodge A (2017) Myanmar energy: the case for smart LNG solutions. Royal Norwegian Embassy of Myanmar
EarthRights International (2008) The human cost of energy: Chevron's continuing role in financing oppression and profiting from human rights abuses in military-ruled Burma (Myanmar). EarthRights International.
Glassman J (2011) The geo-political economy of global production networks. Geogr Compass 5(4):154–164
Gomes L (2020) The dilemma of gas importing and exporting countries. In: OIES papers. Oxford Institute for Energy Studies, Oxford
Hayter R, Barnes TJ, Bradshaw MJ (2003) Relocating resource peripheries to the core of economic geography's theorizing: rationale and agenda. Area 35(1):15–23. https://doi.org/10.1111/1475-4762.00106
Hess M (2004) 'Spatial' relationships? Towards a reconceptualization of embeddedness. Prog Hum Geogr 28(2):165–186
Holliday I (2005) The Yadana syndrome? Big oil and principles of corporate engagement in Myanmar. Asian J Polit Sci 13(2):29–51. https://doi.org/10.1080/02185370508434257
Irarrázaval F (2020) Natural gas production networks: resource making and interfirm dynamics in Peru and Bolivia. Ann Am Assoc Geogr 1–19. https://doi.org/10.1080/24694452.2020.1773231
Jessop B (2007) State power: a strategic relational approach. Polity, Cambridge
Kean T (2017) The Ministry needs to move fast to keep the lights on. In: Frontier Myanmar. https://www.frontiermyanmar.net/en/the-ministry-needs-to-move-fast-to-keep-the-lights-on/. Accessed 4 Sep 2020
Kean T (2018) Does Myanmar's LNG power plan stack up? In: Frontier Myanmar. https://www.frontiermyanmar.net/en/does-myanmars-lng-power-plan-stack-up/. Accessed 4 Sep 2020
Kolås Å (2007) Burma in the balance: the geopolitics of gas. Strat Anal 31(4):625–643. https://doi.org/10.1080/09700160701559318
Kortelainen J, Rannikko P (2015) Positionality switch: remapping resource communities in Russian borderlands. Econ Geogr 91(1):59–82. https://doi.org/10.1111/ecge.12064
Larsen J (1998) Crude investment: the case of the Yadana pipeline in Burma. Bull Concerned Asian Sch 30(3):3–13
Lertchavalitsakul B, Meehan P (2020) Myanmar's contested borderlands. In: Simpson A, Farrelly N (eds) Myanmar: politics, economy and society. Routledge, London
Lindblad T (2015) Foreign direct investment in Indonesia: fifty years of discourse. Bull Indones Econ Stud 51(2):217–237. https://doi.org/10.1080/00074918.2015.1061913
Mackie J (2010) Patrimonialism: the new order and beyond. In: Aspinall E, Fealy G (eds) Soeharto's new order and its legacy: essays in Honour of Harold Crouch, vol 2. ANU Press, Canberra
Mainhardt H (2019) Too high a price for the poor and climate? The World Bank's energy access programme in Myanmar. Bank Information Center Europe, Amsterdam
Mehden FVD, Lewis SW (2006) Liquefied natural gas from Indonesia. In: Victor DG, Jaffe AM, Hayes MH (eds) Natural gas and geopolitics: from 1970 to 2040, pp 91–121. Cambridge University Press, Cambridge
Ministry of Energy and Mineral Resources (ESDM) (2016) The book of electricity statistics number 29
Mol APJ (2011) China's ascent and Africa's environment. Glob Environ Chang 21(3):785–794
Negara S (2016) Indonesia's infrastructure development under the Jokowi administration. Southeast Asian Affairs 1:145–165
Neilson J, Dwiartama A, Fold N, Permadi D (2020) Resource-based industrial policy in an era of global production networks: strategic coupling in the Indonesian cocoa sector. World Dev 135:105045. https://doi.org/10.1016/j.worlddev.2020.105045
Nugraha AT, Prayitno G (2020) Regional disparity in western and eastern Indonesia. Int J Econ Bus Adm (IJEBA) 8(4):101–110

Purwanto WW, Muharam Y, Pratama YW, Hartono D, Soedirman H, Anindhito R (2016) Status and outlook of natural gas industry development in Indonesia. J Nat Gas Sci Eng 29:55–65
Saleh R (2012) Domestic market obligation (DMO) policy and its implementation strategies. Indones Min J 15(1):42–58
Seah SH (2014) Can Indonesia's policy of reconfiguring its energy mix by increasing natural gas usage support its initiatives to reform subsidies? Oxford Institute for Energy Studies, Oxford
Sheppard E (2002) The spaces and times of globalization: place, scale, networks, and positionality. Econ Geogr 78(3):307–330. https://doi.org/10.1111/j.1944-8287.2002.tb00189.x
Simpson A (2016) Energy, governance and security in Thailand and Myanmar (Burma): a critical approach to environmental politics in the South. Ashagate Publishing Ltd., Farmington
Smith A (2015) The state, institutional frameworks and the dynamics of capital in global production networks. Prog Hum Geogr 39(3):290–315
Stokke K, Vakulchuk R, Øverland I (2018) Myanmar: a political economy analysis. Norwegian Institute of International Affairs, Oslo
The Irrawaddy (1997) Burma is now facing its worst economic crisis. The Irrawaddy August 1997. https://www2.irrawaddy.com/article.php?art_id=720. Accessed 8 Apr 2021
Urry J (2013) Societies beyond oil: oil dregs and social futures. Zed Books, London
Wallerstein I (1979) The capitalist world-economy, vol 2 Cambridge University Press, Cambridge
Warburton E (2016) Jokowi and the new developmentalism. Bull Indones Econ Stud 52(3):297–320
Warburton E (2017) Resource nationalism in post-boom Indonesia: the new normal? Lowy Institute for International Policy
Watts M (2004) Resource curse? Governmentality, oil and power in the Niger Delta, Nigeria. Geopolitics 9(1):50–80. https://doi.org/10.1080/14650040412331307832
World Bank (2018) Access to electricity (% of population)—Myanmar. In: World Bank (ed) World development indicators
Yeung H (2016) Strategic coupling: east Asian industrial transformation in the new global economy. Cornell University Press, London
Yusuf A, Sumner A (2015) Growth, poverty, and inequality under Jokowi. Bull Indones Econ Stud 51(3):323–348. https://doi.org/10.1080/00074918.2015.1110685

Alexander Dodge works as an associate professor at the Department of Geography at the Norwegian University of Science and Technology where he also obtained his Ph.D. in 2020. He has previously worked as a postdoctoral research associate at the Department of Geography at Durham University. His research focuses on conceptualizing and accounting for the emergence, transformation, and instability of hydrocarbon production networks in the global economy. He has a particular focus on the role of nation-states in constituting the conditions for stability and change in hydrocarbon production networks. He has previously carried out research on global shifts in liquefied natural gas production networks and the implications for energy markets in Southeast Asia and he is now researching the evolution and ongoing transformation of the UK's strategic position in global oil production networks. He teaches courses in global production networks, innovation studies, and economic geography.

Part II
Scales

Chapter 5
Scale as a Lens to Understand Resource Economies in the Global Periphery

Kirsten Martinus, Julia Loginova, and Thomas Sigler

Abstract Economic geography has generally focussed on industries in, and therefore the economies of, core urban economies. This is problematic when trying to understand how well-established economic processes unfold in global peripheries. This chapter focusses on how the methodologies used at various scales can help better understand the economic processes of global periphery regions by 'zooming in' on the industries that shape such regions, and by 'zooming out' on the processes mediating firm-level activity in peripheries. It takes a closer look at the resource periphery regions of the Pilbara Region in Western Australia, and Northern Komi in North-western Russia, to identify distinct scalar spatial relations that are relevant to understanding how peripheries are contextualised and made. We find similarities in resource economy structures at the regional and sub-national scales, particularly in their socioeconomic situation vis-à-vis their relationship to core national regions, presence of large Indigenous populations, and marginalised political position. However, we find considerable divergence at the national scale. Russian networks are more centralised than Australian networks, where many cities act as key economic hubs. The chapter elucidates the importance of the cross-scale analysis to understand the complexities of resource peripheries.

Keywords Resource peripheries · Global city networks · Mining · Scale · Russia · Australia

K. Martinus (✉)
School of Social Sciences, The University of Western Australia, Perth, Australia
e-mail: kirsten.martinus@uwa.edu.au

J. Loginova · T. Sigler
School of Earth and Environmental Sciences, The University of Queensland, Brisbane, Australia
e-mail: j.loginova@uq.edu.au

T. Sigler
e-mail: t.sigler@uq.edu.au

F. Irarrazaval and M. Arias-Loyola (eds.), *Resource Peripheries in the Global Economy*, Economic Geography, https://doi.org/10.1007/978-3-030-84606-0_5

5.1 Introduction

A growing body of work in economic geography has focussed on global peripheries, especially on territories in which the resources sector plays an important economic role. This chapter aims to use *scale* as a lens to better understand how economic processes of peripheralisation are shaped by a variety of geographical processes through an analysis of resource regions within two nations with very different political economies—Australia and Russia. To do this, we first conceptualise resource peripheries from an economic geography perspective and demonstrate that socio-spatial relationships of resource economies are not uniform across different scales. We discuss the specific resource systems that are central to economic production in the Pilbara (Australia) and northern Komi (Russia). Next, we examine distinct scalar spatial relations between local, sub-national and their respective national jurisdictions by looking at how resources are governed and benefits are redistributed from the resource peripheries to the core. Finally, we examine how Australia and Russia fit within global networks of resources sector firms using corporate data obtained from the Osiris database (Bureau van Dijk 2019).

Through the discussion of similarities and differences in the organisation of resource economy structure between different geographical scales, we demonstrate *how* understanding the processes shaping the relationships of resource peripheries with the global economy depends on the scale of the methodological approach. Using comparative analysis, we argue that understanding how similar outcomes may emerge in very different national settings means drilling down into the nature and motivations of their respective corporations and governments, and the relationships between them.

5.2 Resource Peripheries and Uneven Development

The concept of 'uneven development' is a persistent theme in economic geography that characterises the tendency for capitalist development to reproduce social and economic divisions at multiple scales. Viewed primarily from a neo-Marxian perspective (Smith 2010), uneven development refers to local, regional and global disparities in wages, and access to the means of production more broadly. In other words, capitalism exacerbates, or at least maintains, divisions in relative levels of economic development and with distinct spatial dimensions. This unevenness of development is particularly evident in resource economies in that vast disparities exist between the peripheries where resources are physically located, and the core urban regions in which major corporations process, trade and consume such resources.

Resource peripheries are conceptualised as regions[1] at the intersection of natural resource extraction and the periphery (distance from the resource-hungry economic core). They are often remote regions with difficult to inhabit climates, perceived

[1] We adopt the Australia definition for 'regions' in reference to areas outside of metropolitan cities.

to be ‘sinks’ of raw materials. From the Arctic and sub-Arctic of Canada, Russia and Finland to the Amazon regions of Bolivia, Colombia, Peru, Ecuador and Brazil; from the vast desertscapes of Argentina, Australia, Botswana, Chile and Libya to steppes in Mongolia, the United States and China, resource peripheries are found across all inhabited continents. However, understanding resource peripheries requires a discussion on how the core orchestrates activities carried out in the peripheries, and *how* this has been studied. In this context, we refer to the core as the territories in which key economic activities are organised—being primarily more-developed and more-urban regions. In contrast, peripheries are typically more geographically ‘remote’—being both less-developed and less-urbanised.

Though not formally defined across countries, resource peripheries share several commonalities. Their economies are tied to primary industries such as mining, forestry, oil and gas production, or other extractive sectors, and are dominated by the urban centres housing the bulk of firms, governments and metropolitan population. Resource peripheries are often sparsely populated with a high proportion of Indigenous population, have competing value systems, underdeveloped service economies, and local institutions that are frequently subordinate to national or global capital interests. Residents, particularly those engaging in the more lucrative aspects of resource exploitation are often migrants—both domestic and international—who may in fact be transient or temporarily in a region. In some cases, such ‘territories’ lack full political representation at a national level (e.g. Northern Territory, Australia and Yukon Territory, Canada). More importantly, peripheries are often perceived as ‘backwards’, or at least marginal, in socioeconomic terms vis-à-vis the core. This is despite substantial wealth generated from their natural resource abundance and deep integration in the global economy through international trade and corporate activity. The rapid advent of globalisation has meant the resources industry is highly international in its labour, circuits of knowledge and capital, governance and norms (Tulaeva et al. 2019).

Studies indicate that extraction benefits tend to concentrate outside resource periphery regions, serving multiple scales and limiting opportunities for development in resource peripheries (cf. Phelps et al. 2015; Breul and Revilla Diez 2021; Breul et al. 2019). Economic geography has drawn on the political economic perspective of Innis (Hayter and Barnes 1990), with ‘staples theory’ highlighting the dangers of understanding resource economies through regional models based on ‘conventional’ wisdoms of how urban systems operate (Argent 2013; Martinus 2018). Others have used a value chain approach to show how benefits accrue by core economy investors and manufacturers over local firms and labour (e.g. Atienza et al. 2018; Hanlin and Hanlin 2012).

Resource regions are inextricably linked to power inequalities between resource regions and the cores at global, national or regional scales. How they are integrated into economic networks and value chains is intended to maximise the benefits and minimise the costs of resource extraction for both governments and corporations, being the key actors of the process guiding decision-making across different scales (Irarrázaval 2020a, b). Corporations view remote regions as ‘resource banks’ (Tonts et al. 2013), with infrastructure investment facilitates resource extraction and export with comparatively little returned for peripheral community development. Indeed, regimes throughout history produced and reinforced the centrality of the resource-poor 'core' and sustained underdevelopment in resource-rich periphery, as postulated by Innis's core-periphery model (Hayter and Barnes 1990). This has been observed in emerging economies and resource-rich regions in the Global South (Arias et al. 2014; Bustos-Gallardo 2017; Irarrázaval 2020a, b; Murphy 2012) and advanced capitalist economies resource-rich regions (Argent 2013; Martinus 2018).

The limited ability of resource communities to diversify away from resource extraction is linked to the fact that money earnt in wages often leaves the region through ephemeral workforce arrangements (Martinus 2018). Historically, various exploitative arrangements have been in place to furnish this labour to live in peripheries, including slavery, indentured servitude and forced labour (e.g. Gulags). Now, there are two general systems. First, temporary labour needed in wealthier countries is fulfilled by workers from the global South. For example, Indonesian workers toiling in rubber and palm oil plantations in hot and humid peripheral regions of Malaysia (e.g. Sabah, Sarawak), or South Asian workers labouring in the oil economy of Saudi Arabia's torrid desert regions. Second, in countries with more tightly bound labour regimes, workers are 'imported' to resource peripheries from metropolitan regions. For example, large 'fly-in-fly-out' labour forces in Australia and Chile (Atienza et al. 2020; Martinus 2018; Paredes et al. 2018). The remainder of this chapter will explain the importance of cross-scale examination to more fully understand peripheries.

5.3 Scale as a Lens to Understand Resource Economies in the Global Periphery

Scale is the geographical resolution of global, national, regional or local at which a process or phenomenon is studied. How economic geography understands the uneven (re)distribution of resources across scales of analysis is critical to contextualising resource peripheries (cf. Bridge 2008; Phelps et al. 2015). Two key approaches are of interest to this chapter. First, resource peripheries are conceptualised through related regional models by stressing their context-specific socioeconomic and political dynamics as they sit *outside* of cores (Kortelainen and Rannikko 2015). That is, they are studied as spatially separate in a national system of cities. But, as Hayter et al. (2003) contend, the complexity of resource economies means they should be viewed via multidimensional forces of industrialism (economic dimension), regulationism (political dimension), environmentalism (environmental dimension) and aboriginalism (cultural dimension) found in their unique local contexts. Second, peripheries are studied as an *outcome* of uneven geographies of resource (re)distribution, with multinational corporations (MNCs) and cities of resource economies providing basing points of exchange within a global system of cities. That is, resource peripheries are subject to the forces at play at different scales, including national and global, connecting them to the global economic system through resource dependencies. This is drawn into sharper relief using the relational geography of city networks, connecting resource peripheries to political and economic cores (Loginova 2021; Martinus and Tonts 2015).

Bridging these two conceptual and methodological perspectives enables exploration of interactions between the global, national, sub-national and regional geographic scales (see Bridge 2008; Fleming et al. 2015; Phelps et al. 2015) in the context of global and national dependencies on resource peripheries. At the local scale of the resource periphery, resource dependencies unfold in different ways. Local fortunes are often built off the back of resource 'booms' that drive migration and can enrich resource regions, many of which are—even if ephemerally—some of the wealthiest jurisdictions in their respective countries (Ericsson and Löf 2019). Consider North East Scotland which includes the oil hub of Aberdeen, it has one of the highest gross regional products (GRP) per capita in the United Kingdom outside London. Some peripheries can become the core. For example, San Francisco and Melbourne were relatively small settlements at the time of their respective mid-nineteenth century gold rushes. Many resource peripheries are highly volatile places, characterised by environmental degradation, protests and corruption. Others are either abandoned or depopulated, for example, the many silver and copper mining regions in the American Inter-Mountain West (e.g. Nevada) or the mining regions of northern Spain.

At a sub-national scale of a wider area which includes the local resource peripheries as well as metropolitan centres, resource dependencies have mixed impacts. Some states and provinces' economic development strategies are explicitly tied to resource exploitation. In the United States, the outsized effects of oil and gas in both Alaska and North Dakota have driven both investment and migrants in the past decade. Often this creates quite different subnational outcomes. For example, in the United Arab Emirates (UAE) where Abu Dhabi's gas-rich economy is differently scaffolded to Dubai's tourism and logistics-based development policy. In some cases, single resources companies are highly influential in sub-national politics, for example, the state-owned Jinchuan Group's (JNMC) industrial impact in China's remote Gansu province or the multinational corporation Rio Tinto's impact in Mongolia's South Gobi province tied to the world's largest copper mine (Hatcher 2020; Li et al. 2016).

At a national scale, resource dependencies are most influential when resource peripheries play an outsized role in sustaining national accounts. Oil- and gas-rich countries such as Venezuela, Brunei and Equatorial Guinea are almost entirely reliant on their hydrocarbon exports for revenue, similar to Chile's reliance on copper or Botswana's reliance on diamonds. Resource dependencies generate distinct relationships between domestic resources 'peripheries' and core regions, some of which are redistributive in nature. Examples include Iraq's oil-rich Kurdistan region (Alkadiri 2010), oil-rich Chechnya in Russia (Dunlop and Dunlop 1998) or copper-rich Katanga in the Democratic Republic of the Congo (DRC), in which regional autonomy has at least in part been driven by sizable resource economies. In other cases, national resource exploitation regimes have generated conflict, such as in the Niger Delta Region, where groups have sabotaged oil pipelines and incited violence in reaction to what they perceive to be the state-sanctioned encroachment of multinational firms into regions lacking social and infrastructural investment on a national scale (Hönke 2009; Watts 2004).

At a global scale, resource dependencies tied to resource extraction are perhaps the most extreme. The DRC is one of the most resource-rich countries on Earth, containing vast deposits of copper, gold, uranium, cobalt and coltan—a rare but essential mineral for mobile phone production. The DRC's GDP per capita is approximately US $500, whereas oil-rich Norway's equivalent figure is approximately US $75,000.[2] Such disparities in resource economies occur when the power and control of exploitation rights are concentrated in the hands of few (e.g. Russian oligarchs (Fortescue 2006)) and institutions are maladapted to promote wealth redistribution through wages and taxation. Moreover, resource wealth often has a disruptive impact on democracy and equitable decision-making; in the most recent Transparency International 'Corruption Perceptions Index' (2019), many resource-rich states such as Angola and Equatorial Guinea ranked near the bottom despite being relatively wealthier than peer countries—a clear indication that resource wealth does not translate into positive development outcomes.

Multiple scales define resource peripheries in relation to both resource regimes and the firms that operate within them. Adopting a scalar lens enables us to examine resource peripheries as not only a hierarchical construct (ranging from local and global) but as socially constructed and historically contingent on resource interdependencies. In this analysis, we focus on the cross-scale resource dependencies of Australia and Russia to show the uneven impacts of resource peripherality on the Pilbara and northern Komi Republic. In the sections that follow, we provide a brief background tracing the resource economies of Australia and Russia and then elaborate upon three scalar relationships that connect the regional, subnational, national and global.

5.4 Australia and Russia

Australia and Russia provide the case study settings for this chapter, sharing a number of both similarities and differences. Both are large territorial landmasses whose modern history is tied to imperial expansion. Russia's expansion over the Eurasian landmass preceded Britain's invasion, and subsequent settlement, in Australia by several hundred years, but the outcome of this common history is that both countries have vast 'frontier' regions which are considerable producers of resource wealth (Argent 2013; Bradshaw and Prendergrast 2005).

In the case of Australia, key commodities are iron ore, gold, coal, oil and—more recently—gas. The Australian spatial economic structures bear the hallmarks of the resource-reliant 'settler society'. Its regional economies, such as Pilbara in Western Australia, developed as resource sites within national and global economies driven

[2] https://data.worldbank.org/indicator/NY.GDP.MKTP.CD.

mainly by the global commodity market cycles and the activity of multinational corporations (Argent 2013; Tonts et al. 2013). Russia's immense resource wealth includes oil, gas, coal, iron ore, copper, precious gems and many other minerals and metals (Bradshaw and Connolly 2016). In contrast, the Russian state has historically been strongly involved in the resource industry spatially shaped by the Soviet system of central planning and currently dominated by large state-owned companies (Gazprom and Rosneft) (Bradshaw 2009). Despite decentralisation policies in the 1990s, since the 2000s the government has focused on re-centralisation and direct state involvement in the political and economic affairs in regions, such as the Komi Republic, strengthening the commanding role of Moscow (Bradshaw and Prendergrast 2005).

5.5 Resource Dependencies Between the Regional and the Sub-national Scale

To examine the first scalar relation, we focus on two resource regions, northern Komi Republic in Russia and the Pilbara in Western Australia (WA). Located remotely from the capitals of their respective sub-national areas, Syktyvkar and Perth, these regions are characterised by extreme climate conditions, with northern Komi being one of the coldest, and the Pilbara being one the hottest, inhabited regions on Earth. Northern Komi and Pilbara provide good examples of how resource extraction wealth and regional development are not necessarily linked. Although both regions had large investments in mining-related industrial and transport infrastructure, they are disadvantaged in the development of their towns and in socioeconomic indicators such as economic wealth, public health, education and/or community development in comparison to their metropolitan command centres. Finally, large Aboriginal populations in the Pilbara and ethnic Komi communities are at a power disadvantage in regional decision-making processes and have been historically excluded from resource development. The details of each case below provide some evidence as to how peripherality is differentially experienced and exerted in these two regions, and how despite the vast natural resources of each, development eludes vast swathes of these territories.

5.5.1 Northern Komi

Figure 5.1 demonstrates the location of major projects taking place in sparsely populated areas in taiga and tundra environments. Oil extraction is concentrated around Usinsk, while gas extraction takes place near Ukhta. Metallurgic coal is extracted from deep underground mines near towns of Vorkuta and Ukhta; bauxites are mined north-west of Ukhta. Resource extraction co-exists with semi-nomadic and rural livelihoods of reindeer herding, hunting and cattle ranching in Komi and Nenets communities.

Temporary labour has played a crucial role in the industrial development of the region. Prisoner labour was used to build a railway linking regional capital Syktyvkar with centres of industrial activities including coal mining towns of Vorkuta and Inta

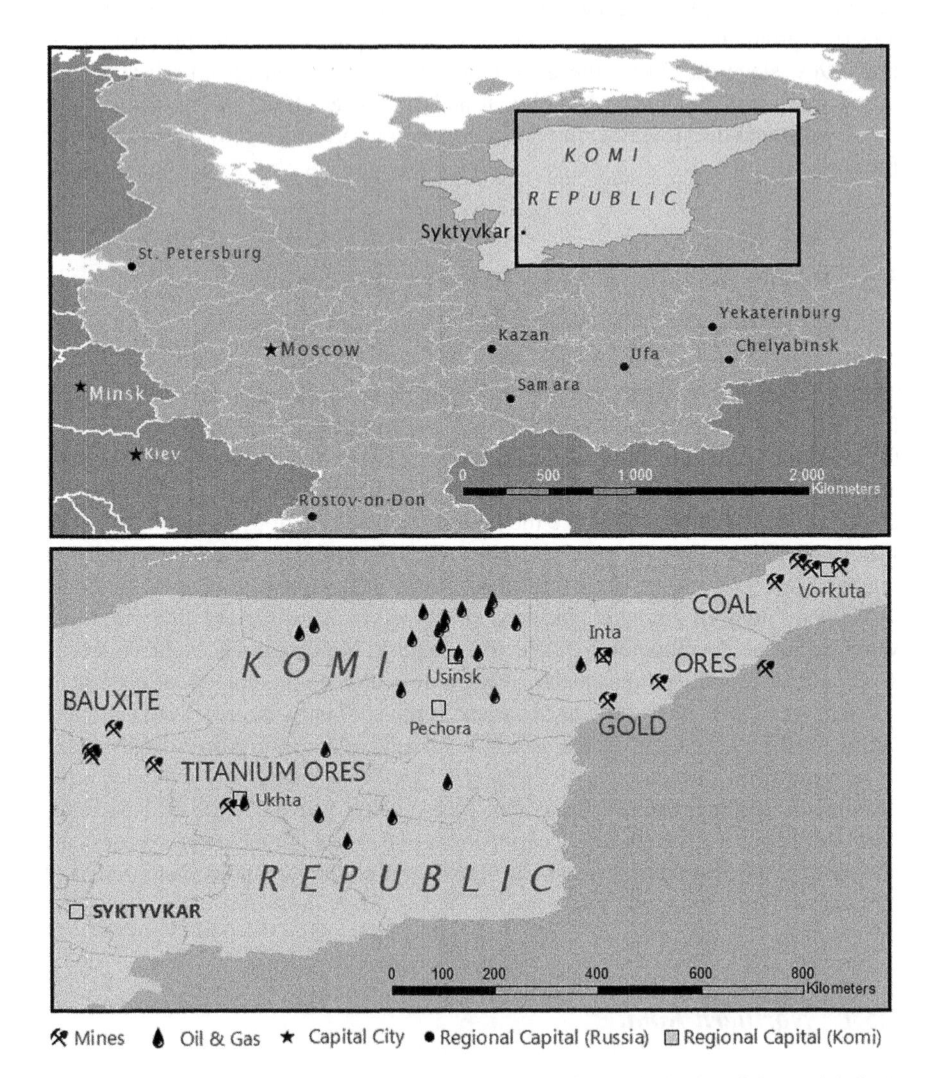

Fig. 5.1 Maps showing the location of Komi Republic within Russia (top) and the spatial distribution of major resource projects in Northern Komi (bottom) (Maps by the authors; Data from the Geoportal of the Komi Republic, the Territorial Fund of Information of the Komi Republic[3])

[3] https://gis.rkomi.ru/

established between the 1930s and 1950s as sites of Stalin-era Gulags (forced labour camps). Usinsk is a centrally planned oil town built in the 1970s accommodating 50,000 people who moved from all over the Soviet Union to northern Komi following the Soviet government directives to develop the northern resource region. Since then, the population in these towns has significantly declined (up to 50%), with a mixture of the long-term workforce residing locally and long-distance commuters (termed 'vakhtoviki') staying in workers' camps. Formal statistics on temporary labour within Russia are lacking, however, estimates point at significant intra- and interregional flows connecting northern regions with large urban centres (Eilmsteiner-Saxinger 2011). Currently, about 25,000 residents of the Komi Republic are employed in the resources sector, or 6.7% of total regional employment (KomiStat 2020). This is disproportionally low when compared to the share of the oil and gas industry of Northern Komi (40%) in the gross regional product (KomiStat 2020). Companies benefit from relying on temporary labour that provides access to mobile and qualified workers who prefer to live in warmer regions and more liveable cities; however, this resource dependence disadvantages the local and regional economy.

Resource dependencies have increased with the move from the state-regulated to capitalist industrialism. The oil and gas industry has been rapidly developing since the 1970s, with production and infrastructure owned and managed by the state. Lukoil, one of the largest Russian oil corporations based in Moscow (11% of Russia's proved oil reserved, 15% of crude oil production and 15% of crude oil refining (LukoilDataBook 2019)), acquired the assets of the Komitek oil company in 1999 controlled at that time by the Komi Republic's Government. Lukoil has been the largest oil producer in Komi, and gradually integrated Komi oil into the national and international networks of refining, transportation, and petroleum product marketing networks. Lukoil-Komi, its subsidiary based in Usinsk, is responsible for sustaining the extraction of oil and negotiating benefits with the regional and local governments. The benefits include direct contributions to local and regional development in the form of modernisation of kindergartens, schools, and hospitals as well as support to regional cultural and sporting events through social partnership agreements (Wilson and Istomin 2019). This support, although minimal, is significant for local and rural municipalities with otherwise limited resources for development that reach remote regions from the Komi Republic's level of governance.

However, people and local governments have been increasingly concerned with the operations of large corporate structures, which have been notorious for lack of transparency over their operations and revenues. In recent years, local populations have increasingly expressed their dissent with resource extraction by Lukoil-Komi due to public concerns over water pollution with oil spilled from ruptured pipelines and lack of community participation (Loginova and Wilson 2020). Demands for responsibility for the local environment have been linked to calls for the greater participation of Indigenous people (Komi and Nenets) in regional resource development (Loginova and Wilson 2020). Although companies stimulated local participation in labour force, there are concerns that such employment contributes to crowding out of traditional livelihoods and resource dependency poses barriers for diversification of regional economy.

5.5.2 *Pilbara Region*

The Pilbara is one of the most mineral-rich regions on Earth, having the world's largest iron ore deposits (Regional Development of Australia 2014), in addition to ample deposits of gold, spudomene (lithium), nickel, alumina, copper and other minerals. Australia is also the second-largest liquefied natural gas (LNG) exporter in the world (behind Qatar), with WA contributing over half of this—much of which is from the Pilbara. Figure 5.2 illustrates the range of mineral resources, with its gas reserves lying primarily offshore.

In 2018, the Pilbara produced more than 70% of WA mineral and energy production, worth almost 70 billion Australian dollars annually. The region had a population of 62,093 people and 63,850 jobs, of which an estimated additional 50,000 workers (or 78% of the jobs) were flown in (Remplan 2020). Administratively, it is comprised of four large local government areas (LGAs), whose populations are mostly in the largest city of each. The two major settlements of Port Hedland and Karratha contain approximately half of the region's permanent population (ABS 2016), the rest of which is spread out over more than 500,000 km^2. LGAs are responsible for their own governance (specifically roads, waste removal, planning) and some socioeconomic development, with the Pilbara Redevelopment Commission working on behalf of all to broaden the regional economy, ensure standards of government services and develop infrastructure.

The Government of the State of Western Australia administers the *Royalties for Regions* scheme, giving money extracted from the regions back to the regions (WA DRD 2016). This scheme injects money back into regional communities for capacity building, guaranteeing that 25% of all royalties earned from mining would be spent in non-metropolitan areas. Introduced in 2008 after recognition that limited to no money from the regions was being reinvested there, it has been wrought with controversy, with the $A1 billion per annum accused of being used for political pork-barrelling (Kagi 2018; Tonts et al. 2013). Whilst there are local content policies in place, the complexity of what is 'local' for firms largely housed in Perth, lack of regional industrial capacity and complex tendering practices mean limited money is spent on regional businesses (Parliament of Australia 2018). The WA State Government is responsible for education, hospitals, roads and other major infrastructure, owns resource royalties, environmental governance and issuing of mining leases. The State capital city of Perth has a population of around 2 million (ABS 2016), and therefore substantially more amenity, infrastructure, government and business than anywhere in the Pilbara.

Whilst there is a resident population in the Pilbara, its labour force largely adheres to long-distance arrangements on 2–6 work week rosters. The bulk of these are termed

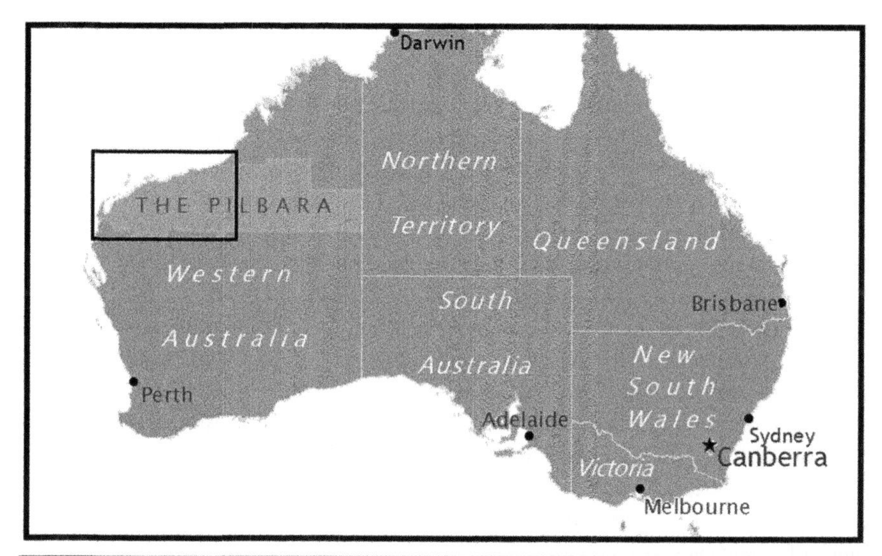

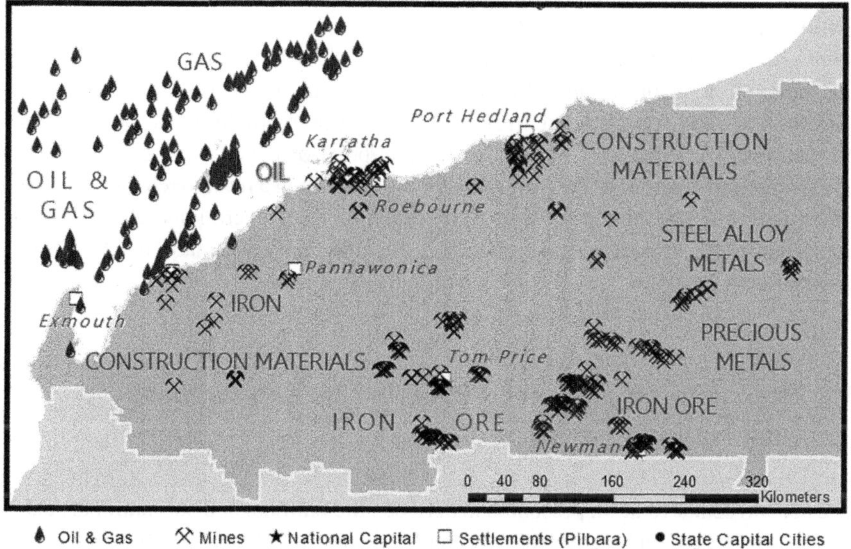

Fig. 5.2 Maps showing the location of the Pilbara within Australia (top) and the spatial distribution of major resource projects (bottom). (Maps by the authors; Data from Western Australia Department of Mines, Industry Regulation and Safety[4])

[4] http://www.dmp.wa.gov.au/Mines-and-mineral-deposits-1502.aspx.

fly-in/fly-out (FIFO) workers, coming largely from Perth or a handful of coastal high-amenity towns. Workers are usually housed in mining camps rather than nearby towns, and provided with food and basic services by resource companies and their subcontractors. The FIFO system is highly cost-effective and well-established, with little time for leisure during work weeks. Wages are largely spent outside of resource towns and back in worker source communities. This means almost complete leakage of monies out of resource regions with companies taking profits, the government taking royalties and workers taking wages elsewhere (Martinus 2018).

The region's Indigenous population has increasingly been incorporated into discussions with the State Government as the traditional landowners of Western Australia, forcing mining companies to negotiate Native Title Agreements (Austrade 2015), which are voluntary agreements between native title parties and others wishing to use or manage areas owned by Indigenous peoples as part of the National Reconciliation Action Plan (National Native Title Tribunal 2020). In the Pilbara, this is with the Yamatji Marlpa Aboriginal Corporation. Final statutory decisions are made by the High Court of Western Australia, but often go into long appeals given the significance of what is at stake on both sides (Bunch 2020). The Federal Government is also involved through the role of the Department of Environment to protect the environment and Indigenous sacred sites (Siewart 2020). Rio Tinto's recent destruction of the Juukan Gorge, an irreplaceable world heritage site of cultural significance for at least 46,000 years, points to the irreverence that large international corporations and government often show to local Indigenous populations. The decision has precipitated structural changes to the firm's leadership, including a push to relocate its global headquarters from London to Perth to be closer to daily operations (Harvey 2020), the firing of top executives and enquiries into the decisions made by the Federal Environment Minister.

5.6 Resource Dependencies Between the Sub-national and National Scales

The second scalar relation presented is between the national and sub-national scales. Sub-national geographical territories can be formally defined—such as states and provinces. Both Australia and Russia are federations. However, the centralisation of power is much more significant in Russia, as states in Australia have larger autonomy (Loginovà et al. 2020). Below we consider this in the context of resource dependencies of Komi Republic and the Pilbara with their respective national and State governments.

5.6.1 Komi Republic

The Komi Republic was established in early Soviet Russia based on the rights of minorities to self-determination. During Stalin's period of Russification, the republics and their titular nationalities were strictly controlled by Moscow. The Constitution was adopted in 1994 granting regional autonomy in the development and implementation of socioeconomic policies and sovereignty of natural resources. However, over the last two decades the scope of regional competencies has been gradually reduced in favour of federal regulations, specifically in relation to natural resources. Currently, the regional authorities and people are powerless actors in terms of influencing the allocation of licences and distribution of benefits (with the exception of programmes of Corporate Social Responsibility which are negotiated between the companies and regional or local authorities).

The hierarchical resource dependencies between Komi Republic and the Federal Government are evident in the tax collection schemes and fiscal regional equalisation. Taxes on the subsoil resources are collected on the federal level contributing to a large share of federal budget (46% in 2019 (MinFin 2021)). Komi Republic is just one of the 'donor' regions. These are then redistributed among poorer agricultural regions of Russia through fiscal equalisation policies. The federal collection of the subsoil tax significantly limits opportunities to capture economic benefits from the oil and gas industry (Fjærtoft 2015) which is the largest industry in Komi Republic, contributing around 40% to the regional gross product (KomiStat 2020). The attempts to establish a regional sovereign wealth fund that would capture resource benefits were unsuccessful due to low and inconsistent volumes of the inputs and disagreements with the federal authorities.

More recent disagreements between the regional and national priorities occurred over the closure of borders that accompanied the COVID-19 pandemic. In June 2020, temporary workers' camps in northern Komi experienced some of the largest concentrations of infections in the region, raising significant concerns among the wider population over their exposure to the virus through shift workers (TASS 2020). Regional authorities while recognising the risks did not restrict or limit flows of workers entering and leaving the region for shift work and justified the decision by the importance of the oil sector for the economic resilience of country and the region. The COVID-19 pandemic hit the Russian hydrocarbon sector particularly hard, with negative ripple effects for its resource peripheries.

5.6.2 Western Australia

The economic significance of the Pilbara is only understood by examining industry production and the movement of labour across the nation, as well as the governance arrangements by the State and National governments. In Australia, mining has the highest value add to the national economy (ABS 2019), along with other industry sectors closely related to mining such as financial services, construction, and professional scientific and technical. The significance of the mining sector to national economic wealth is well acknowledged: the Commonwealth Government attributing Australia's breaking of world records for the 'longest run of uninterrupted growth in the developed world' to the resource sector and specifically in sustaining Australia through the global crisis of 2007/10 (Department of Industry, Science, Energy and Resources 2018).

Emerging as an outcome of Federation in 1901 where each separate Australian British colony united to form the Commonwealth Government. The State governments have large autonomy and control over the operation of their states, being granted partial self-governance by the British in 1860s. As such, the Commonwealth provides services for national benefit (environmental protection, health care, social security, defence, border protection), receiving incomes from the states through various taxes—including goods and services tax (GST)—which is redistributed amongst the states through fiscal equalisation policies. Issues such as state development, education, health care and state border control are all under State jurisdiction. Both State and Commonwealth governments play roles in foreign affairs and export matters in the context of their specific jurisdictions. The contribution of Western Australian resources is extremely significant—not only in the direct export sale of resources itself, but also in the indirect demand for services across the nation. Even during the COVID-19 pandemic, Western Australia was the only Australian state to deliver a budget surplus in 2020 due to higher than expected iron ore prices and royalties (Shine 2020).

There are two recent examples of how this State power plays out in relation to the nation, and the way in which resource peripheries are landscapes of contestation. Firstly, for the first time in its history WA closed its borders fully to other states during the COVID-19 pandemic to protect its residents and business. This created a national backlash as billionaire Clive Palmer (from Queensland) sued the State for A$30 billion in lost revenue. His arguments centred on the income losses of his Balmoral iron ore project, and the inability for FIFO workers from other states to enter. He claimed border closures were unconstitutional, forcing the Commonwealth to act in the reopening WA borders. Whilst the Prime Minister of Australia initially warned the Premier of WA would lose the court case (Bell and Hamlyn 2020), three days later the Commonwealth withdrew its legal challenge acknowledging the State was acting in the best interests of its citizens and industry (Carmody 2020). There were no challenges to other states closing borders. The case highlights the degree of autonomy of Western Australia, and how relative control over resource wealth sits at a State scale.

The second example of the power relationships played out in the resource sector by the WA State government, the Commonwealth and corporations relates to the Rio Tinto—Juukan Gorge case discussed above. Investigations of the destruction of this significant Aboriginal cultural site revealed that Rio Tinto had been pressuring the Commonwealth to urgently hand over environmental approval powers to the State of Western Australia before the completion of a major enquiry into national environmental laws. The move was made primarily to have mine construction underway before more stringent rules came in place, given that the Commonwealth report found Australia's biodiversity and natural environments were in an 'unsustainable decline' and the WA State government was seen as being more sympathetic to the needs of the mining industry (Cox 2020). Rio Tinto currently holds 1,780 approvals to destroy sacred Aboriginal sites (Allam 2020). The example highlights how corporations and governments negotiate exploitation and extraction of the resource peripheries, while populations in regional areas (in this instance Indigenous populations) are often powerless to shape more positive and equitable outcomes.

5.7 Resource Dependencies Between the National and Global Scales

The third scalar relationship is between the national and global scales showing the *different* ways that resource corporations of the two nations globalise—one reflecting globally oriented capitalism, the other more akin to state-led capitalism. We operationalise this in terms of global network relations of firms located in Australian and Russian cities and towns. Numerous scholars have pointed out not only the value in understanding urbanisation as a product of 'world city networks', but increasingly in understanding how smaller, and more peripheral, cities fit within these networks (Phelps et al. 2015, 2018; Scholvin et al. 2019). Outside of well-known 'global cities', many places are peripheral to global economic systems, either due to their geographical peripherality, low overall levels of global economic activity, or economic systems that focus activity toward national, rather than international networks.

To illustrate how global-national scalar relations can differ across contexts, we provide a characterisation of the role of Australia and Russia within resource corporate networks at the global and national scales. Corporate networks show connections of firms operating in the two countries as headquarters or subsidiaries of multinational and domestic resource firms. We used data from the Osiris database (Bureau van Dijk 2019) that provides detailed information about multinational firms listed on stock exchanges. Firms were included in the analysis if either its head office or subsidiary was classified as 'mining and quarrying' according to the European industry classification code NACE. This includes largely upstream and downstream activities and specialised service providers. We constructed a 'two mode' network, because we begin with the links between two firms (based on the headquarter-subsidiary relation) and transform the link into a tie between two locations of these offices. The end result is thus a network graph with places as nodes that are connected through resource-sector linkages, with focus on resource firms with offices in Australia and Russia, respectively.

Much of the previous work in this vein has applied network analysis to better understand how national production systems are structured through cities. Loginova et al. (2020) demonstrate variegated corporate city network structures in Australia and Russia showing the different ways that oil and gas corporations of each nation globalise: the Australian network reflects a decentralised global system, whereas the Russian system is clearly more centralised on Moscow. Figure 5.3 illustrates this variegation by showing maps of corporate networks of resource (oil, gas and mining) firms linking cities in Australia and Russia to the rest of the world.

The network graph shows cities that are important for the resource activity at the national scale. The Australian network appears to be nationally decentralised with Perth ($n = 1{,}249$ connections), Sydney ($n = 746$), Melbourne ($n = 573$), Brisbane ($n = 446$) and Adelaide ($n = 215$) having the largest number of connections. International connections span all five other inhabited continents, connecting Australian state capital cities into the global network, with the most connected cities being

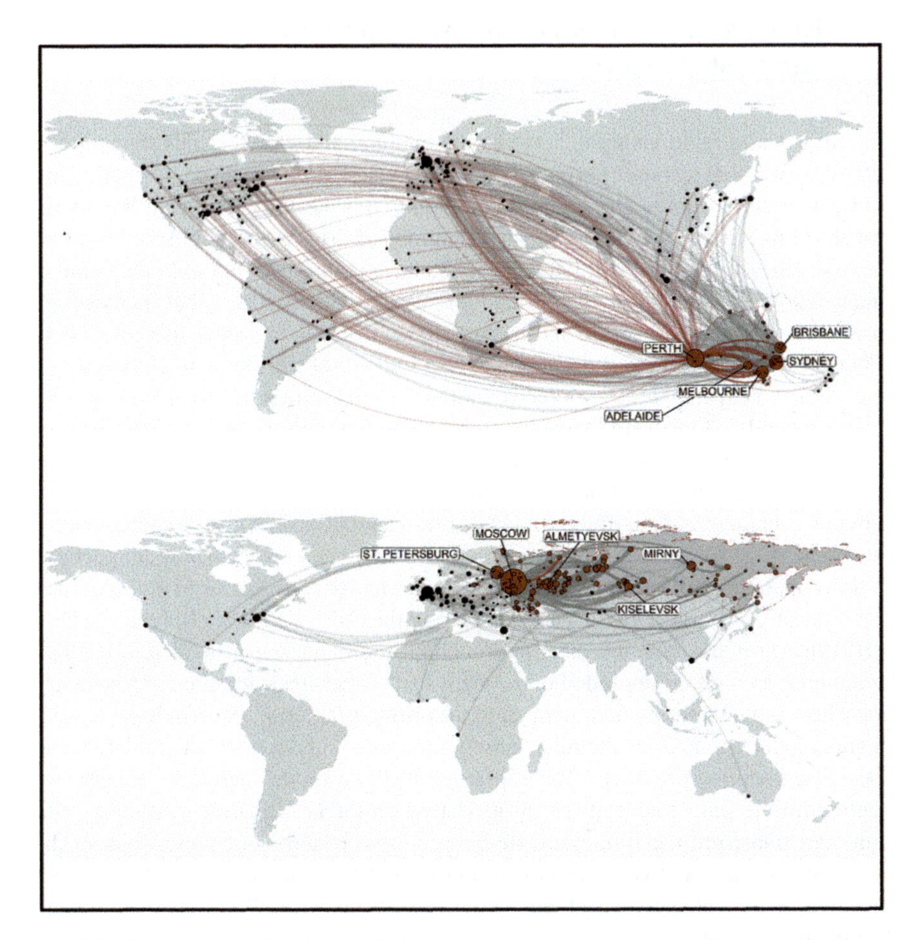

Fig. 5.3 Map of multinational firm relationships in the resources sector of Australia (top) and Russia (bottom). Names are displayed for five nodes (cities) with the highest number of firm offices. Network corporate relationships with more than one connections are shown only. Edges of Western Australia and Komi Republic are coloured in red. (Maps by the authors; Data from Bureau van Dijk (2019))

London ($n = 430$), Houston, New York, Singapore, Tokyo, Johannesburg, etc. The high international and domestic connectivity of Perth places Western Australia in the core of the Australia's resources industry.

The Russian network is highly centralised around Moscow (the capital and by far the largest city hosting majority of headquarters of resources firms, $n = 573$) with extensive intra-national linkages across 190 cities. This is followed by Saint Petersburg ($n = 114$), Almetyevsk ($n = 85$) and Mirny ($n = 49$). Among foreign cities, London is the most connected ($n = 106$), followed by Nicosia ($n = 29$), Amsterdam (22), Vienna ($n = 21$) and a few others each with less than 10 connections. Despite large-scale resource extraction in the Komi Republic, it is Moscow that is at the core of Russia's resources network.

These city networks of firms show the key metropolitan areas concentrating resource firm activity (resource hubs). The divergent geographies of the two nations demonstrate how truly globalised the Australian resources sector is, and how reliant the Russian system is on central decision-making in Moscow. This spatial distribution of the resource firm ownership structures characterises the expression of spatial inequality in the two countries at the national and global scales (see Loginova et al. (2020) for further details). Revenues generated from the extraction and export of natural resources are associated with firms headquartered in specific metropolitan areas nationally and internationally, that therefore benefit disproportionally from resource activity in remote regions. The accounts of specific firm activity and their contribution to regional development in Western Australia and Komi Republic detailed in the previous section provide understanding as to why Perth and Moscow are the most dominant in their respective networks.

5.8 Conclusions

Resource peripheries are deeply integrated into global economic processes. A better understanding of resource peripheries can be achieved by adopting geographical approaches that uncover multidimensional processes of peripheralisation at multiple scales—from the local to the global. It is impossible to understand the complexities of resource peripheries if studied at one scale. Our case studies from Russia and Australia show how resource peripheries are places that are and have historically been characterised by uneven power relationships across multiple scales due to a lack of economic and political autonomy at the local scale and the resource dependence of core economies at other scales (Hayter et al. 2003). In the Pilbara and northern Komi, local interests are sidelined or marginalised for global or national ones, producing conflict and vulnerability due to the volatile nature of commodity markets or the geopolitical significance of resources.

This is less true at the sub-national scale, at which Western Australia holds considerable autonomy over both resource development and the proceeds thereof. The Komi Republic, on the other hand, lacks autonomy in many regards, and is largely an administrative territory (rather than a sovereign land). At the same time, considerable divergence exists at the national scale, where Moscow plays a central role in the peripheralisation of Russia's northern regions, such as Komi Republic, whereas in Australia corporate activity is spatially decentralised, diffused and rather global in scope. This suggests that resource interdependencies between Australia and global markets, while the Russian system ensures that both decision-making and profit, are centralised and that international relationships are centrally mediated.

Recognising that all peripheries similar but also unique, a key outcome from this chapter is that these areas peripheries are 'made' through strategic decisions driven by corporate activity about the re-allocation of resources and re-distribution of benefits in the context of resource extraction regimes. In other words, corporate strategies, national economic policies, geopolitical relationships as well as international processes (e.g. driven by demand in global commodity markets) effect resource peripheries. Resource 'peripheries' are only peripheral insofar as they are politically marginalised or socio-economically disadvantaged, and in an ideal scenario become less socio-economically peripherialised and disadvantaged over time. This may be through investment in social services, local content policies and economic diversification programmes, such as investing in alternative industries such as sustainable eco-tourism around natural or Indigenous assets given recent critiques of the predatory and unsustainable nature of the industry (Bianchi and de Man 2021; MacNeill and Wozniak 2018; Pezzullo 2007). Such an approach is critical to facilitate a shift from being disadvantaged dependent places within the global economy to sites of less dependent economies, empowered local population, greater support for Indigenous customs and protected natural environments.

References

ABS (2016) The census of population and housing. Australian Bureau of Statistics, Canberra

ABS (2019) Australian National Accounts: national income, expenditure and product. ABS Cat. No. 5206. Australian Bureau of Statistics, Canberra

Alkadiri R (2010) Oil and the question of federalism in Iraq. Int Aff 86(6):1315–1328

Allam L (2020) Rio Tinto still has 1,780 approvals to destroy Aboriginal scared sites, Jookan Gorge inquiry told. The Guardian. https://www.theguardian.com/australia-news/2020/oct/16/rio-tinto-still-has-1780-approvals-to-destroy-aboriginal-sacred-sites-juukan-gorge-inquiry-told

Argent N (2013) Reinterpreting core and periphery in Australia's mineral and energy resources boom: an Innisian perspective on the Pilbara. Aust Geogr 44(3):323–340

Arias M, Atienza M, Cademartori J (2014) Large mining enterprises and regional development in Chile: between the enclave and cluster. J Econ Geogr 14(1):73–95

Atienza M, Arias-Loyola M, Lufin M (2020) Building a case for regional local content policy: the hollowing out of mining regions in Chile. The Extractive Industries and Society 7(2):292–301

Atienza M, Lufin M, Soto J (2018) Mining linkages in the Chilean copper supply network and regional economic development. Resour Policy 70:101154

Austrade (2015) Western Australia: information on engaging with traditional owners. https://www.austrade.gov.au/land-tenure/Engagement-guide/western-australia-information-on-engaging-with-traditional-owners

Bell F, Hamlyn C (2020) Clive Palmer 'highly likely' to win WA coronavirus border closure legal fight, Prime Minister warns. ABC News. https://www.abc.net.au/news/2020-07-29/clive-palmer-highly-likely-to-win-wa-border-challenge-pm-says/12501872
Bianchi RV, de Man F (2021) Tourism, inclusive growth and decent work: a political economy critique. J Sustain Tour 29(2/3):353–371
Bradshaw M, Connolly R (2016) Russia's natural resources in the world economy: history, review and reassessment. Eurasian Geogr Econ 57(6):700–726
Bradshaw M, Prendergrast J (2005) The Russian heartland revisited: an assessment of Russia's transformation. Eurasian Geogr Econ 46(2):83–122
Bradshaw M (2009) The Kremlin, national champions and the international oil companies: the political economy of the Russian oil and gas industry. Geopolit Energy 3
Breul M, Revilla Diez J (2021) "One thing leads to another", but where?—gateway cities and the geography of production linkages. Growth Chang 52(1):29–47
Breul M, Revilla Diez J, Sambodo MT (2019) Filtering strategic coupling: territorial intermediaries in oil and gas global production networks in Southeast Asia. Journal of Economic Geography 19(4):829–851
Bridge G (2008) Global production networks and the extractive sector: governing resource based development. J Econ Geogr 8(3):389–419
Bunch A (2020) High Court rejects FMG's native title appeal over Pilbara land. WAToday. https://www.watoday.com.au/national/western-australia/high-court-rejects-fmg-s-native-title-appeal-over-pilbara-land-20200529-p54xv4.html
Bureau van Dijk's Osiris database (2019) https://www.bvdinfo.com/en-gb/our-products/data/international/osiris
Bustos-Gallardo B (2017) The post 2008 Chilean Salmon industry: an example of an enclave economy. Geogr J 183(2):152–163
Carmody J (2020) Commonwealth withdraws from Clive Palmer border case, Prime Minister's letter to WA Premier reveals. ABC News. https://www.abc.net.au/news/2020-08-02/government-removes-support-for-clive-palmers-push-to-open-border/12515948
Cox L (2020) Letter reveals Rio Tinto urged transfer of powers to WA ahead of environment law review. The Guardian. https://www.theguardian.com/environment/2020/oct/02/rio-tinto-made-early-call-for-morrison-to-transfer-environmental-approval-powers-to-wa
Department of Industry, Science, Energy and Resources (2018) The Australian resources sector—significance and opportunities. https://www.industry.gov.au/data-and-publications/australias-national-resources-statement/the-australian-resources-sector-significance-and-opportunities
Bureau van Dijk (2019) Commonwealth withdraws from Clive Palmer border case. Prime Minister's letter to WAPremier reveals. ABC News. https://www.abc.net.au/news/2020-08-02/government-removes-support-for-clive-palmers-push-to-open-border/12515948
Dunlop JB, Dunlop JB (1998) Russia confronts Chechnya: roots of a separatist conflict. Cambridge University Press, Cambridge
Eilmsteiner-Saxinger G (2011) 'We feed the nation': benefits and challenges of simultaneous use of resident and long-distance commuting labour in Russia's Northern hydrocarbon industry. J Contemp Issues Bus Gov 17(1):53
Ericsson M, Löf O (2019) Mining's contribution to national economies between 1996 and 2016. Miner Econ 32(2):223–250
Fjærtoft DB (2015) Modeling Russian regional economic ripple effects of the oil and gas industry: case study of the republic of Komi. Reg Res Russ 5(2):109–121
Fleming D, Measham T, Paredes D (2015) Understanding the resource curse (or blessing) across national and regional scales: theory, empirical challenges and an application. Aust J Agric Resour Econ 59(4):624–639
Fortescue S (2006) Russia's oil barons and metal magnates: oligarchs and the state in transition. Palgrave Macmillan, London
Hanlin R, Hanlin C (2012) The view from below: 'lock-in' and local procurement in the African gold mining sector. Resour Policy 37(4):468–474

Harvey B (2020) Mark McGowan urges Rio Tinto to relocate head office from London to Perth. The West Australian. https://thewest.com.au/business/mining/mark-mcgowan-urges-rio-tinto-to-relocate-head-office-from-london-to-perth-ng-b881667603z
Hatcher P (2020) Global norm domestication and selective compliance: the case of Mongolia's Oyu Tolgoi mine. Environ Policy Gov 30(5):252–262
Hayter R, Barnes T (1990) Innis' staple theory, exports, and recession: British Columbia, 1981–86. Econ Geogr 66(2):156–173
Hayter R, Barnes T, Bradshaw M (2003) Relocating resource peripheries to the core of economic geography's theorising rationale and agenda. Area 35(1):15–23
Hönke J (2009) Transnational pockets of territoriality: governing the security of extraction in Katanga (DRC). Leipziger Univ.-Verlag
Irarrázaval F (2020a) Contesting uneven development: the political geography of natural gas rents in Peru and Bolivia. Polit Geogr 79:102161
Irarrázaval F (2020b) Natural gas production networks: resource making and interfirm dynamics in Peru and Bolivia. Ann Am Assoc Geogr 111(2):1–19
Kagi J (2018) Langoulant inquiry into WA finances slams Barnett government decisions. ABC News. https://www.abc.net.au/news/2018-02-20/langoulant-inquiry-into-wa-finances-slams-barnett-government/9465312
KomiStat (2020) Territorial authority of the federal service of state statistics in the Komi Republic. https://komi.gks.ru/
Kortelainen J, Rannikko P (2015) Positionality switch: remapping resource communities in Russian borderlands. Econ Geogr 91(1):59–82
Li Y, Beeton RJ, Sigler T, Halog A (2016) Modelling the transition toward urban sustainability: a case study of the industrial city of Jinchang, China. J Clean Prod 134:22–30
Loginova J, Sigler T, Martinus K, Tonts M (2020) Spatial differentiation of variegated capitalisms: a comparative analysis of Russian and Australian oil and gas corporate city networks. Econ Geogr 96(5):422–448
Loginova J, Wilson E (2020) "Our consent was taken for granted": a relational justice perspective on the participation of Komi people in oil development in northern Russia. In: Johnstone RL, Hansen AM (eds) Regulation of extractive industries: community engagement in the arctic. Routledge, Milton
Loginova J (2021) City networks binding resource peripheries to economic and political cores: a Northeast Asian perspective. Asian Geogr 1–20
LukoilDataBook (2019) LUKOIL in Russia and the world. In: Lukoil data book. https://www.lukoil.com/InvestorAndShareholderCenter/ReportsAndPresentations/AnnualReports_info
MacNeill T, Wozniak D (2018) The economic, social, and environmental impacts of cruise tourism. Tour Manage 66:387–404
Martinus K, Tonts M (2015) Powering the world city system: energy industry networks and interurban connectivity. Environ Plan A 47(7):1502–1520
Martinus K (2018) Inequality and regional development in advanced capitalist economies. Geogr Compass 12(11):e12405
MinFin (2021) Annual information on the execution of the federal budget. https://minfin.gov.ru/ru/statistics/fedbud/
Murphy J (2012) Global production networks, relational proximity, and the sociospatial dynamics of market internationalization in Bolivia's wood products sector. Ann Assoc Am Geogr 102(1):208–233
National Native Title Tribunal (2020) About indigenous land use agreements (ILUAs). [Webpage]. http://www.nntt.gov.au/ILUAs/Pages/default.aspx
Paredes D, Soto J, Fleming DA (2018) Wage compensation for fly-in/fly-out and drive-in/drive-out commuters. Pap Reg Sci 97(4):1337–1353
Parliament of Australia (2018) Keep it in the regions. Commonwealth of Australia, Canberra
Pezzullo PC (2007) Toxic tourism: Rhetorics of pollution, travel, and environmental justice. University of Alabama Press, Tusaloosa

Phelps N, Atienza M, Arias M (2015) Encore for the enclave: the changing nature of the industry enclave with illustrations from the mining industry in Chile. Econ Geogr 91(2):119–146
Phelps N, Atienza M, Arias M (2018) An invitation to the dark side of economic geography. Environ Plan A 50(1):236–244
Regional Development of Australia (2014) Resources and beyond. Regional Development of Australia, Canberra
Remplan (2020) Economy, jobs and business insights: the Pilbara. WA Department of Primary Industries and Regional Development, Perth. https://app.remplan.com.au/pilbararegion/economy/trends/population?state=Kp9MIN!6Y46i16mKFVMWJlCxnv9kiQtwhXlrFwhohotghaw5
Scholvin S, Breul M, Revilla Diez J (2019) Revisiting gateway cities: connecting hubs in global networks to their hinterlands. Urban Geogr 40(9):1291–1309
Shine R (2020) WA to deliver surplus budget despite pandemic payments on back of iron ore price rise. ABC News. https://www.abc.net.au/news/2020-09-25/surplus-budget-in-wa-despite-coronavirus-pandemic-payments/12705148
Siewart R (2020) Rio Tinto must sack boss and front environmental committee: greens. https://rachel-siewert.greensmps.org.au/articles/rio-tinto-must-sack-boss-and-front-environment-committee-greens
Smith N (2010) Uneven development: nature, capital, and the production of space. University of Georgia Press
TASS (2020) *Новый Очаг Коронавируса Выявили Среди Вахтовиков в Усинске в* Коми (New coronavirus hotspot detected among shift workers in Usinsk in Komi). https://tass.ru/obschestvo/8835735
Tonts M, Martinus K, Plummer P (2013) Regional development, redistribution and the extraction of mineral resources: the Western Australian Goldfields as a resource bank. Appl Geogr 45:365–374
Transparency International (2019) Corruption perceptions index. https://www.transparency.org/en/cpi
Tulaeva SA, Tysiachniouk MS, Henry LA, Horowitz LS (2019) Globalising extraction and Indigenous rights in the Russian arctic: the enduring role of the state in natural resource governance. Resources 8(4):179
Watts M (2004) Resource curse? Governmentality, oil and power in the Niger Delta, Nigeria. Geopolitics 9(1):50–80
Western Australian Department of Regional Development (DRD) (2016) Regional Development Strategy 2016–2025. WA DRD, Perth
Wilson E, Istomin K (2019) Beads and trinkets? Stakeholder perspectives on benefit-sharing and corporate responsibility in a Russian oil province. Eur Asia Stud 71(8):1285–1313

Kirsten Martinus works as an urban and economic geographer at The University of Western Australia. She holds Bachelor in Economics (Hons) from The University of Western Australia, and a Ph.D. in Urban and Regional Planning from Curtin University. Her research focuses on the links between resource wealth, regional innovation and global competitiveness, balancing fundamental research with applied outputs to inform strategic policy and decision-making for local and state government. Her current projects examine industry hubs in renewable energy, technological disruption in industry, employment distribution in cities and the impact of technology.

Julia Loginova has completed her Ph.D. at The University of Melbourne, and is currently a Postdoctoral Research Fellow in Economic Geography at The University of Queensland in Australia. She received her graduate degree in Economics and Law and previously worked in regional development in Russia. Julia's main research interests are within economic geography and political ecology focussing on the globalisation of resource firms, geographies of energy transitions, the role of the state and collective action. Julia's dissertation focused on oil and gas extraction regions in northern Russia and interrogated multi-scale challenges of subnational resource governance and issues of community participation.

Thomas Sigler is an urban and economic geographer at The University of Queensland. He holds a Bachelor of Arts in Geography and International Relations from the University of Southern California, and Master of Science and Ph.D. in Geography from the Pennsylvania State University. His teaching, supervision and research are broadly focussed on economic change in global cities, and the impacts of globalisation on cities. His current projects focus on global firm networks, the urban impacts of the sharing economy, and housing & land use change.

Chapter 6
Reproducing the Resource Periphery: Resource Regionalism in the European Union

Juha Kotilainen

Abstract The chapter explores the reproduction of resource peripheries through a study on the raw materials policy of the European Union (EU). Established in the late 2000s, the raw materials policy has been mainly focused on non-renewable non-energy minerals. The policy has included the explicit goals of increasing the extraction of minerals in the EU territory and to secure the supply of raw materials from sources outside the EU. Paying attention to the spatial divisions these policy goals are creating, the chapter critically examines the EU's raw materials policy as a multi-scalar undertaking, in which the main players are the European Commission, governmental organisations of the EU member states, industry interest groups, and the various administrative bodies of the European regions at a scale below the nation-state. There are political and economic actors in the regions who support the intensification of resource extraction as a part of their attempts to deal with the social and economic conditions in the regions. As a result, this agency produces and reproduces the regional identities as resource-based. The chapter specifically explores the ways in which such resource regionalism plays a role in the reproduction of resource peripheries.

Keywords Minerals policy · Extractivism · Resource nationalism · Resource regionalism · Resource peripheries · European Union

6.1 Introduction

The spatial scales that mostly have been explored in terms of resource extraction are defined by local communities that are both the targets for adverse impacts of the extraction and sources of workers for the resource industries (Kotilainen 2018); nation-state governmental policies, with ownership of sub-soil minerals and politics of resource nationalism (Veltmeyer & Petras 2014; Johnson & Ericsson 2015); and multinational corporations, for which findings generally suggest them to have

J. Kotilainen (✉)
Department of Geographical and Historical Studies, University of Eastern Finland, Joensuu, Finland
e-mail: juha.kotilainen@uef.fi

F. Irarrazaval and M. Arias-Loyola (eds.), *Resource Peripheries in the Global Economy*, Economic Geography, https://doi.org/10.1007/978-3-030-84606-0_6

conflictual relations with the local communities and collaborative relations with the nation-state governments (Bridge 2008; Arboleda 2020; Irarrázaval 2021). The reach of the companies is global, while at the same time limited by politics at the national scale. In addition to these scales, there is a need to investigate the roles of supra-national actors above the nation-state, such as the European Union (EU), and regions below the nation-state, in order to complement the picture over global resource extraction. Consequently, there are questions of how resource policies that are practised across these spatial scales seek to overcome the problem of allocation of land for the extraction of minerals, considering the challenges of land use often being locked for other purposes in densely populated spaces and, related to this, of the relations of resource policies for the reproduction of resource peripheries.

In this chapter the focus is on the question of how spatial divisions of production, which lead to the emergence and re-emergence of resource peripheries, are produced through social and political processes. The idea that there are spatial cores and peripheries implies that peripheries provide raw materials to the core spaces so that value is added into the products in the cores (Ciccantell & Patten 2016; Schaffartzik et al. 2016). But how exactly do resource peripheries emerge and re-emerge in relation to the core spaces? How do regions at a smaller scale than the nation-state reproduce their positions in relation to core spaces? Who are the agents for the production of resource peripheries? The argument that is forwarded in this chapter is that at least a partial answer can be found by investigating development policies concerning regions at a scale between the national and the local, as well as policies at the regional scale addressing resources to be exploited. This policy formulation occurs in a multi-scalar context, which is also explored in this chapter. It is suggested that peripheries are not only produced from the outside of the peripheral spaces, but that there are actors within the regions, mainly those focussed on the economic development of a region, whose agency is crucially important in the processes of reproduction of the resource periphery.

The chapter investigates the reproduction of resource peripheries by examining the formation of the raw materials policy of the EU as well as its impacts spatially. Established in the late 2000s, the raw materials policy of the EU has mainly been focused on non-renewable non-energy minerals. It has been an explicit goal of the policy to increase the extraction of minerals in the territory of the EU and to secure the supply of raw materials from sources outside the EU. Paying attention to the spatial divisions these policy goals are creating, the chapter critically examines the EU's raw materials policy as a multi-scalar undertaking, in which the main players are the European Commission, governmental organisations of the EU member states, industry interest groups and the various administrative bodies of the European regions at a scale below the nation-state.

Region is understood here as the scale of administration and spatial organisation between the national state and local communities. While there have been transformations and upsurges in mineral policies in some of the member-states of the EU, such as Sweden and Finland (Johnson & Ericsson 2015; Haikola & Anshelm 2016), it is also possible to identify political and economic actors in some of the European regions who support the intensification of resource extraction as a part of their attempts to improve the often adverse social and economic conditions resulting from industrial decline. As a result, this agency reproduces the regional identities as resource-based. The chapter specifically explores the ways in which such resource regionalism plays a role in the reproduction of resource peripheries. Therefore, in this chapter resource periphery is seen to be actively and interactively produced by agents across several spatial scales: national, supra-national and regional below the nation-state, which is suggested to be a novel contribution to the literature on resource peripheries.

6.2 Global Extractivism, Resource Nationalism and Resource Regionalism in a Multi-Scalar Perspective

There are two sets of broader scholarly debates addressing the societal drivers and impacts of the exploitation of natural resources that are of interest here. In critical research literature, the first debate is framed as a concern over the rise of extractivism or new extraction, which are notions that have been especially used in relation to the social, economic and environmental impacts from the extraction of resources in Latin America (Veltmeyer & Petras 2014; Dougherty 2016). The other debate is over resource nationalism (Haslam & Heidrich 2016; Kaup & Gellert 2017; Wilson 2015; Koch & Perreault 2019), where the key focus is on the emphasis that national governments put on securing financial returns from the natural resources extracted within the territories they govern. The positions of new extraction and resource nationalism are related, as both aim at explaining the reasons for and consequences of the increase in the exploitation of resources. In both, the national scale is in a crucial position, which has also been criticized more generally from a methodological perspective (Arboleda 2020). Yet extractivism has been less tied to the national scale, and its proponents see more clearly there to be overexploitation of resources, both nature and human. Both extractivism and resource nationalism are theoretical perspectives that have been used for developing economies, but the question is to what extent they would be applicable in a European context and at other spatial scales than the nation-state (Humphreys Bebbington & Bebbington 2010; Irarrazaval 2020). The central arguments put forth in the literature on both extractivism and resource nationalism are explored next.

There are large numbers of global linkages related to resource extraction through production chains and networks (Bridge 2008; Schaffartzik et al. 2016). In this setting, some spaces globally tend to become sources for minerals, and, accordingly, resource peripheries at a global scale have been examined through critical perspectives on 'extractivism' (e.g., Veltmeyer & Petras 2014) or 'new extraction' (Dougherty 2016). Extractivism points to political choices that lead to joint efforts by corporations and the state on where and how resources are extracted, as well as dependence on resource extraction of the local, regional and national economies.

Extractivism has been recognized as a phenomenon at the global scale (Schaffartzik et al. 2016; Arboleda 2020) although it has been most prominently used for Latin America (Veltmeyer & Petras 2014; Dougherty 2016). As a summary of various studies, extractivism has been concluded to mean that the extractivist sectors contribute a large share of total value added, the export of primary commodities is amongst the most important sources of revenue, and labour and natural resources are exploited beyond their ability to reproduce themselves (Schaffartzik et al. 2016).

On the other hand, the literature on extractivism or new extraction has pointed out that governments at the national scale aim at creating economic benefits for their societies by supporting the exploitation of natural resources in their territories. Societies related to the new extraction have been characterised as post-neoliberal (Veltmeyer & Petras 2014), meaning that they practice policies at the national scale that are announced to be socially inclusive, and this is sought to be reached by increasing the activities of resource extraction. All this has been understood to derive from the cycles of capitalism and the needs for resource extraction that it produces.

Resource nationalism, on the other hand, pays attention to how nation-states develop their economies on the use of natural resources and at the same time they wish to protect their own right for the exploitation of the resources within their territories. Resource nationalism is generally understood to be combined of political aspirations guiding economic principles through which economic benefits from the exploitation of natural resources within the territory of a nation-state are kept to that nation (Nem Singh 2014). Haslam and Heidrich (2016) include three categories of actions under resource nationalism: the maximization of public revenue; the assertion of strategic state control, that is, ability to set a political or strategic direction to the development of the extraction of resources; and enhancement of developmental spillovers from extractive activity. Wilson (2015) argued that while economic dynamics function as an enabling factor for resource nationalism to emerge, political institutions are an equally important conditioning factor shaping resource nationalism. Therefore, resource nationalism can be defined as a strategy where governments use economic policies to improve returns from the resource industries.

The reason for national governments for promoting resource nationalism is that they do not see that liberal economic policies would lead to natural resources within the territory of a nation-state to be developed in ways that will offer maximum benefits for the host state, which is why national governments prefer to define the terms for resource exploitation in ways that advance national goals. Three different versions of resource nationalism have been identified: rentier, developmental and market-based (Wilson 2015). There are also specific strategies that are deployed as means towards executing resource nationalism, including policies that aim at local or state ownership of resource industries; policies constraining the operations of resource firms through industrial policy requirements and trade regimes; and policies that are designed to capture economic rents for public purposes through taxation.

Following Wilson's (2015) argument, we can state that while the theories on resource nationalism point out that there are various means for reaching the goal of increasing returns from the extraction of resources, it is important to understand the politics around how and why certain decisions are made. The means are one thing, but the political decisions on the choice of exact means another. Kaup and Gellert (2017), as well, seek to combine the economic and political dimensions of resource nationalism, as they understand resource nationalism to be an action by state actors in extractive peripheries to gain economically and politically. They see resource nationalism as a cyclical process, shaped by the strategies of hegemonic powers and their challengers (Andreucci 2017).

Nation-state governments are key players in resource nationalism. While the functions of the state are represented at several spatial scales, the policies of a nation-state government are the most obvious example of the role that the state plays in providing the socio-political conditions for the extraction of resources. However, the state operates at other spatial scales as well, and it is present at the local scale in the form of local public administration, but also, and in the case of policies on resources currently, at a supra-national scale, such as the EU, which has institutions in place to design and implement public policies. In addition to the local and supra-national scales, the state is present at the regional scale as well through the organs and institutions of the public administration in the regional centres of power.

Throughout these spatial scales, the state has an important role in the extraction of resources. The state is the main provider of knowledge on the deposits of minerals (Kama 2020; Irarrázaval 2021). The state is chiefly responsible for the accumulation of geological knowledge on the locations of minerals and the quality of the ore grades, and it assumes this role in tandem with the institutions of science by collecting the information and storing it in the archives of state bureaucracies and administrations (Braun 2000). Far from being politically neutral, this scientific knowledge is a crucial part of the political economies that nation-states have been creating for centuries. The state is the apparatus that has created the hubs that contain information on mineral wealth. This information is, subject to the conditions dictated by each nation-state, utilised by companies when they are making plans for the exploration of deposits of minerals. It is characteristic for minerals that nation-states have withheld the rights to the knowledge on them, and, in addition, they have withheld the rights on how that knowledge is to be distributed to potential utilisers of those resources and raw materials.

The linkage of nation-states and the rights on information on mineral resources is tight, as is illustrated by the fact that within the EU the principles of permitting in mineral extraction have not been harmonised but kept under control by the individual national governments (European Commission 2013). Consequently, the state has a role in providing the conditions for companies to carry out their activities within the territory of a state jurisdiction. Similarly to the role of the state in global production networks (Smith 2014), in terms of the extraction of resources, the state has the role of directing the investments and actions that are related to the material acts of removing the minerals from the subsoil. There is variation in the extent to which this knowledge is freely available to private businesses. The availability may depend on the specific minerals in question and their strategic importance for the national governments.

In other theoretical contexts, the regional scale has been argued to have certain importance in the production that draws on the vicinity of natural resources (Kinnear

& Ogden 2014; Rehner et al. 2014). For the regional scale, the political can be seen to have much importance. Although at the regional scale similar means as at the national scale for accumulating returns from the resource industries do not exist, the regional scale can be a highly politicised issue. Resource regionalism as understood here can be seen as political action at the regional scale that aims at maximising the economic benefits from the exploitation of natural resources at the regional scale.

The following will explore recent developments regarding the role of the state in the EU across various spatial scales in its participation in the extraction of resources through its raw material policies. Importantly, these policies are results of politics across several spatial scales. The EU represents a unique supra-national scale of state organisation. However, as the occurrences in the EU take place at various scales simultaneously, there is, consequently, a significant cross-scale aspect to the way minerals extraction is dealt with. The spatial scales of importance in what follows are the EU, the member states or nation-states within the EU, as well as the regions within the territories of the member states, forwarding their own policies on the extraction of resources. The interest in this chapter is especially in exploring how the making of these cross-scale policies lead to political support for resource regionalism on one hand and how the regions themselves act as agents of these political-economic processes. The overall result is the reproduction of the position of certain spaces as resource peripheries, that is, spaces that provide the existing political-economic system with commodities to be exported to other spaces (the cores) where they are refined and consumed.

6.3 Raw Materials Policy of the European Union

There are concerns that started to emerge in the EU by the turn of the millennia on the raw material supply for the manufacturing industries being at risk (Tiess 2010; European Commission 2013; Kotilainen 2021). While there have been spaces in Europe that have specialised in the production of primary commodities, such as metals or coal (Radkau 2008; Avango et al. 2019; Kivinen et al. 2020), over time the extraction of minerals in Europe has waned despite the continuous demand for these commodities for manufacturing and energy production. As a consequence, imports started to continuously exceed the exports of raw materials, and the industries dependent on the supply of raw materials together with the Commission of the European Union started paying attention to the biased situation with the balance of trade in commodities in the 2000s (European Commission 2008).

The broad outcome of these worries has been an increase in considerations by actors in the EU administration as well as in business organisations on how to expand the frontiers of resource extraction, either by reusing materials or by seeking to discover and utilise new deposits. This has led to attempts by some departments of the European Commission together with industry organisations and national governments of the EU member states to prepare more detailed strategies that emphasise the need, in addition to the recycling and recovery of metals, to increase the mining of minerals from primary deposits and secure a smooth trade in these minerals for the European industries, from European and extra-European sources alike. This policy field is called here raw materials policy of the EU, and it can be seen to consist of various political documents and incentives for action by various political actors within the EU. The justification for the new raw materials policy drew on the perceived importance of raw materials for manufacturing in the EU and, consequently, its economy (European Commission 2008; European Commission 2017). The concerns over the uninterrupted supply of minerals have been related to two complementary developments. First, industry-oriented sections of the public administration and industry representatives equally have been ready to take action towards securing a continuous supply of raw materials for their industries. Second, viewing the same issue from a different perspective, there have been governments and regional policy and development actors who have seen revitalisation of mining activities as a way to improve employment figures and thereby help local and regional economies that are struggling with structural problems of their national, regional and local economies to reach a growth trajectory.

Consequently, a tangible political focus on raw materials started to emerge in the 2000s as a new policy field of the EU (European Commission 2008; Tiess 2010, 2011). Generally, policy making in the EU is constituted of a complex web of institutions and actors (Rittberger & Winzen 2015; Wonka 2015). Two major institutions in the EU politics at the scale of the union, the European Parliament and the European Commission, can be considered to have significant political power as they prepare EU legislation (Thomson 2015). The European Parliament is an institution of representative democracy and has legislative power (Rittberger & Winzen 2015), while the European Commission prepares European-scale policies and legislations (Wonka 2015). Formal policy documents, communications, prepared by the European Commission, targeted at other EU institutions and the member-states of the EU, have been a crucial tool to forwarding the raw materials policy. A few vital policy documents by the European Commission in the field of raw materials policy have defined the aims for what has been termed the Raw Materials Initiative (European Commission 2008, 2013).

The European Commission is split vertically into political and administrative power on one hand and horizontally into sectoral departments called Directorate-Generals (Wonka 2015). The most relevant Directorate-General regarding the raw materials policy has been a department specialising on the internal market, industry, entrepreneurship and small and medium sized enterprises. In addition, the department responsible for the environment has had a notable secondary role because it has been necessary to define the relations of the extraction of resources to the protection of nature within the EU (European Commission 2011).

Within the EU politics, the governments of the member states are active political actors in their own right, trying to make the EU policies move towards directions that best suit the political aspirations of each member state, while the European Commission has a coordinating role towards the member states (Rittberger & Winzen 2015; Wonka 2015). Moreover, there is a spatial scale of great political and ideological significance within the EU. This scale is that of the regions, existing at smaller scales than the nation-state. However, regions appear in multiple forms administratively and in relation to the national scale (Keating et al. 2015). Nevertheless, regions have significant decision-making power in the distribution of funds within the EU. Like the member-states, the regions also seek to promote their own ends by engaging in politics especially towards the European Commission and by creating alliances with other regions with similar political interests. Finally, lobbying carried out by interest groups is a common way of doing politics in the EU (Mazey & Richardson 2015). Industry associations are amongst the lobbying groups, and in terms of the raw materials policy their input has been important.

The raw materials policy of the EU can be seen to have had four outcomes (Kotilainen 2021). First, the Raw Materials Initiative has been formulated. It was launched in 2008 (European Commission 2008; Thiess 2010), and it is the most important part of the raw materials policy, since it has an impact on the other actions within this policy field. In principle, the Raw Materials Initiative also covers other raw materials than those acquired by mining, that is, timber, but minerals are clearly its most important aspect (Tiess 2010). The Raw Materials Initiative sets three pillars on which the raw materials policy of the EU rests (European Commission 2008). The first is focussed on ensuring a fair and sustainable supply of raw materials from global markets, which means securing the import of minerals from outside Europe. The second pillar places emphasis on ensuring a sustainable supply of commodities from mineral deposits within the EU. Thirdly, the pillar that concentrates on boosting resource efficiency and increasing the amount of recycling aims to improve recycling and recovery. The aim is, therefore, to secure a steady supply of raw materials for industries in the EU from sources within the EU, non-EU countries, and by recycling and recovery. Despite the last pillar, a significant part of the raw materials policy of the EU is focussed on mining of commodities from primary sources, and the extraction of minerals has been expected to occur, first, on the territory of the EU, and second, in countries outside the EU from where the commodities would be traded to the industries in Europe.

As a second outcome of the raw materials policy of the EU, a list of critical raw materials has been drawn up three times. This list is a result of analysis of raw materials along two axes: one that estimates the economic importance of each raw material, and a second that focusses on the risk of shortage in its supply (European Commission 2008; 2014a; 2017). The list was prepared first in 2011, and in updated form in 2014 and 2017.[1] It demonstrates the wide scope of the raw materials policy that these lists in their different versions have included a large variety of different minerals in terms of their purpose of use by different industries.

A third outcome of the raw materials policy has been the minerals policies of the individual member states. As is emphasised within the Raw Materials Initiative, it is one of the major policy goals of this political initiative communicated by the European

[1] The list published in 2017 included antimony, beryllium, borates, cobalt, coking coal, fluorspar, gallium, germanium, indium, magnesium, natural graphite, niobium, phosphate rock, silicon metal, tungsten, platinum group metals, light rare earths and heavy rare earths, baryte, bismuth, hafnium, helium, natural rubber, phosphorus, scandium, tantalum, and vanadium (European Commission 2017).

Commission to other political institutions of the EU to intensify mining within the EU (European Commission 2008). The most general goal is that each member state should develop their own minerals policy, yet as the European Commission has no direct power to enforce its policies over the member states in the issue of exploitation of minerals, the goal is stated in the documents not as an order but as a wish. In order to proceed towards the goal of increasing mining in Europe, the European Commission has urged the EU member states to take several tangible steps towards an intensified minerals policy (European Commission 2013; European Commission 2014b). The aim has been that by unifying policies and practices across the member states, rather than creating a patchwork of different member-state specific mineral policies, this would lead to a complete minerals policy at the scale of the EU.

There are member states that have already been developing their own mineral policies, and the European Commission has been seeking to create a comprehensive picture of these developments (European Commission 2010). Currently the EU could not have a minerals policy without the involvement of its member-states, because mining is an activity that is in the hands of the individual nation-states, as they each have their own systems, principles, legislations and procedures for granting permissions for mining activities (European Commission 2010). In addition, land use planning, which has importance from the perspective of mining activities by allocating land to various purposes, is, as well, governed separately by the planning systems of each nation-state.

There is variation as to which member states have been active in the process of designing their own policies, and where the focus of the policies are (European Commission 2014b), and some governments of the member states have been more eager than others to promote the intensification of mining in their territories. Other governmental minerals policies tend to put more emphasis on recycling and recovery. Some governmental policies also express more concern over the supply of raw materials for the manufacturing industries, while others are more prone for promoting mining as such. In the latter case, mining has been seen to provide an export product with commodities that despite their low degree of processing nevertheless would provide the local and regional economies, often in declining regions, with employment.

Finally, at a scale below the nation-state, a new resource regionalism has been emerging in Europe. The fourth outcome of the raw materials policy has been an intensified focus at the scale of European regions on the extraction of minerals, as well as intensified collaboration between regions that identify themselves as mining regions. This is an important aspect of the minerals policy, even if it has been developing as a side effect to the more general raw materials policy. Regions have also been active in participating in crafting the overall minerals policy by designing minerals policies of their own and formed alliances of mining regions across the EU.

6.4 The Role of Regions in Reproducing the Resource Periphery

As illustrated above, it is a policy goal of the European Commission to increase the production of primary commodities within the EU. The raw materials policy addresses a need to allocate more land for mining and consider the location of deposits in spatial planning (Kotilainen 2021). The question that follows is how could the policy goal be realised in practice, as existing land use limits the possibilities for the extraction of minerals and underground mining technologies are a source of risk due to potential instability and contamination (Goldthau & Sovacool 2016; Bomberg 2017), and, consequently, in many densely populated spaces the extraction of minerals would be untenable. Hence, the extraction could be carried out in the less populated rural spaces. This, in turn, leads to consideration of the role of European regions in the implementation of the raw materials policy. Region is understood here as the scale of administration and spatial organization between the national state and local communities. Politically, the idea of the region has been crucial in the construction of the EU (Keating et al. 2015); regions have been seen to have relevance for the economy and politics in a Europe whereby the role of nation-states would diminish. Regions are also delegated with administrative powers, and they have powers to distribute funds allocated to the regional scale from the European Commission. There are usually multiple actors with operational powers in a region, including regional administrations and development agencies.

In this context, there are certain regions that have become active in promoting policies that are part of the larger raw materials policy. The regions can be seen to have the following roles in organising resource extraction. In the first place, typically for EU policy making, there are varieties of regional initiatives and alliances that have been emerging. First, there are individual regional pro-extraction policies. Individual European regions with extractive policies and strategies have had linkages to governmental mineral policies that have been designed at the national scale and have had similar goals of increasing or securing the extraction of minerals. However, the regional policies are not necessarily derived from the national ones in the sense that they would be governed from the nation-state governmental organs; these regional minerals policies have been much more independent (Avango et al. 2019).

Second, while it is common that regions seek to build cooperation within the EU, and such collaboration between regions is encouraged and supported by the European Commission, this time these linkages have been emerging on the topic of extraction of minerals and closely related industrial production. Consequently, alliances of regions to bring together initiatives on increased resource extraction have been mushrooming in Europe, often co-financed by the European Commission. The alliances of regions include cross-scale collaboration in the form of industrial and regional development projects between the regional alliances and the European Commission. Such alliances have included ‘Mining and Metallurgy Regions of EU’ (European Commission 2021; Mireu 2021); ‘Smart and Green Mining Regions of the EU’ (European

Union 2021b; Remix 2021), and 'European Network of Mining Regions' (European Union 2021a). This networking of regional actors on the topic of extraction of minerals is a phenomenon at the European scale. It has involved regions below the nation-state and from a diverse range of EU member states, including Austria, Czech Republic, Finland, Germany, Greece, Poland, Portugal, Slovakia, Spain, Sweden and United Kingdom (as a former EU member state). Included have been regions such as Castilla y Leon (Spain), Lower Silesia (Poland), Lapland (Finland), Västerbotten (Sweden), Cornwall (UK), Košice (Slovakia), Styria (Austria) and Saxony (Germany). It is the aim of the policies regionally to provide conditions for increased resource extraction in the regions. In order to support this goal and also to cut opposition, buzzwords related to ideas such as "greenness" have also been used. The individual regional policies and the creation of alliances across the continent show a tendency towards active regional ideas for increased resource extraction in the regions involved.

The question that follows is: What motivates the regional actors to carry out these actions and policies? Several factors can be identified. First, there is the history of a region that has seen it specializing on the exploitation of nature's resources. What characterises the regions participating in these activities, is that they have a strong legacy of resource exploitation (Avango et al. 2019; Kivinen et al. 2020). The raw material policies, at the regional scale but also at the larger scales, reproduce this role in the spatial division of labour for the regions. Second, the regions involved often are ones with characteristics of a struggle to deal with the impacts of declining regional and local economies on employment and aging population structures (Kotilainen et al. 2015). This has followed major restructuring in national and international systems of production, and, consequently, the regions have been under on-going structural restructuring of their economies. This is possibly linked with a downward trend in the number of population in a region as well as aging of the population there. Third, there is perceived lack of suitable alternatives to create employment within the region. Fourth, aspirations of individual persons in the field of regional development and politics can be argued to have a significant role in some cases (Halonen 2019).

The contribution of the exploration of such political-economic developments at the regional scale and across regions to the understandings of how resource peripheries re-emerge is twofold. First, these are processes through which the identities of the actors within the regions become tied to the idea of them being focused on the extraction of resources and closely related industrial production such as steel production, a raw material itself to be used in manufacturing possibly elsewhere. Second, these policy initiatives seek to establish linkages from these regions, where the raw materials are extracted, to other regions where they are refined, manufactured, value added to them, and consumed.

As an idea, resource regionalism in the European case can be seen to be produced as a cross-scalar process. Actors, most significantly development agencies, at the regional scale participate as agents in this process. However, resource regionalism is not produced at the regional scale only. It is a target for political action originating at other spatial scales. Political decision-making at the scale of the EU is significant in the production of resource regionalism as an idea that points to the centrality of regions in creating the conditions for the production of raw materials and in creating the incentives and conditions for regions to thrive within the politics of raw material extraction. Moreover, within the multi-scalar framework focused on the extraction of raw materials and creation of resource regions, also the governmental bodies at the national scale have importance through the ways they contribute to the formulation of national minerals policies, thus in turn encouraging the regions to move towards resource regionalism.

6.5 Conclusions

One of the lessons to be learned from this examination on the design and implementation of raw materials policy in the EU is that the question of what constitutes a resource periphery and where such peripheries might lie needs fine tuning. At the same time, as there have been strong political pressures to design and implement a policy that would support the increased extraction of minerals within the EU, these political initiatives have contributed to a reproduction of spatial divisions within Europe. In turn, this has been leading to spaces where the minerals are extracted and those spaces where they are utilised. Crucially, actors at the regional scale have been active in these processes. As the raw materials policy at the European scale is leading to a division of spaces into those extracting the minerals and those utilising the extracted commodities, resource peripheries are being produced within the political discourse as a multi-scalar process. Resource regionalism plays a crucial role in this process; it is both an incentive for actors at the regional scale in some European regions to push the regions towards intensifying the extraction of resources, and, at the same time, it is a result of policies at multiple scales, including the regions themselves, but also the national scale and the supra-national scale of the EU.

Overall, it continues to be an implicit but important idea within the raw materials policy of the EU that the less populated spaces provide other spaces with raw materials into which value is added there. Spaces of extraction emerge and re-emerge in relation to these other spaces through the policies and politics of extraction (Kotilainen 2021). Actors in the regions are active in reproducing their positions in relation to these spaces. It is an explicit aim of the regions participating in the extraction initiatives that they seek to promote increased activities by corporations and smaller scale enterprises operating in the extractive industries. By activating the policies of resource extraction, the regions are, in effect, reproducing their role as resource periphery in comparison to other regions. However, resource regionalism can be seen as an idea that is produced in a multi-scalar process. Ultimately, the two processes of emergence of resource regionalism and reproduction of resource peripheries are related. In the European context, it is the process of emergence of resource regionalism that is significantly (re)producing the resource periphery.

References

Andreucci D (2017) Resources, regulation and the state: Struggles over gas extraction and passive revolution in Evo Morales's Bolivia. Geoforum 61:170–180

Arboleda M (2020) Planetary mine: territories of extraction under late capitalism. Verso, London & New York

Avango D, Kunnas J, Pettersson M, Pettersson, Ö, Roberts P, Solbär L, Warde P, Wråkberg U (2019) Constructing northern Fennoscandia as a mining region. In: Keskitalo ECH (ed) The politics of arctic resources. Change and continuity in the "Old North" of Northern Europe. Routledge, Abingdon & New York

Bomberg E (2017) Shale we drill? Discourse dynamics in UK fracking debates. J Environ Plann Policy Manag 19(1):72–88

Braun B (2000) Producing vertical territory: geology and governmentality in late Victorian Canada. Ecumene 7(1):7–46

Bridge G (2008) Global production networks and the extractive sector: governing resource-based development. J Econ Geogr 8(3):389–419

Ciccantell PS, Patten D (2016) The new extractivism, raw materialism and twenty-first century mining in Latin America. In: Deonandan K, Dougherty ML (eds) Mining in Latin America. Critical approaches to the new extraction. Routledge, Abingdon

Dougherty ML (2016) From global peripheries to the earth's core: the extraction in Latin America. In: Deonandan K, Dougherty ML (eds) Mining in Latin America. Critical approaches to the new extraction. Routledge, Abingdon

European Commission (2008) The raw materials initiative—meeting our critical needs for growth and jobs in Europe. Communication from the Commission to the European Parliament and the Council. COM (2008) 699 final, Brussels 4.11.2008

European Commission (2010) Improving framework conditions for extracting minerals for the EU. Exchanging best practice on land use planning, permitting and geological knowledge. European Commission, Enterprise and Industry

European Commission (2011) Non-energy mineral extraction and Natura 2000. Guidance document. Publications Office of the European Union, Luxembourg

European Commission (2013) On the Implementation of the Raw Materials Initiative. Report from the Commission to the European Parliament, the Council, the European Economic and Social Committee and the Committee of the Regions. COM (2013) 442 final, Brussels 24.6.2013

European Commission (2014a) Report on critical raw materials for the EU. Report of the ad hoc Working Group on defining critical raw materials. May 2014

European Commission (2014b) Report on national minerals policy indicators. Framework conditions for the sustainable supply of raw materials in the EU. European Commission, Enterprise and Industry Directorate-General. Brussels, February 2014

European Commission (2017) On the 2017 list of Critical Raw Materials for the EU. Communication from the Commission to the European Parliament, the Council, the European Economic and Social Committee and the Committee of the Regions. COM (2017) 490 final. Brussels 13.9.2017

European Commission (2021) Mining and metallurgy regions of EU, https://cordis.europa.eu/project/id/776811. Accessed 16 Feb 2021

European Union (2021a) European network of mining regions, https://keep.eu/projects/268/European-Network-of-Mining-Re-EN/. Accessed 16 Feb 2021

European Union (2021b) Smart and green mining regions of EU, https://keep.eu/projects/18853/Smart-and-Green-Mining-Regi-EN/. Accessed 16 February 2021

Goldthau A, Sovacool BK (2016) Energy technology, politics, and interpretative frames: shale gas fracking in Eastern Europe. Global Environ Polit 16(4):50–69

Haikola S, Anshelm J (2016) Mineral policy at a crossroads? Critical reflections on the challenges with expanding Sweden's mining sector. Extract Ind Soc 3:508–516

Halonen M (2019) The long-term adaptation of a resource periphery as narrated by local policy-makers in Lieksa. Fennia—Int J Geogr 197(1):40–57

Haslam PA, Heidrich P (2016) From neoliberalism to resource nationalism: States, firms and development. In: Haslam PA, Heidrich P (eds.) The political economy of natural resources and development: from neoliberalism to resource nationalism. Routledge, Abingdon
Bebbington DH, Bebbington AJ (2010) Extraction, territory, and inequalities: gas in the Bolivian Chaco. Canad J Develop Stud 30(1–2):259–280
Irarrazaval F (2020) Contesting uneven development: the political geography of natural gas rents in Peru and Bolivia. Polit Geogr 79:102161
Irarrázaval F (2021) Natural gas production networks: resource making and interfirm dynamics in Peru and Bolivia. Ann Am Assoc Geogr 111(2):540–558
Johnson EL, Ericsson M (2015) State ownership and control of minerals and mines in Sweden and Finland. Miner Econ 28:23–36
Kama K (2020) Resource-making controversies: Knowledge, anticipatory politics and economization of unconventional fossil fuels. Prog Hum Geogr 44(2):333–356
Kaup BZ, Gellert PK (2017) Cycles of resource nationalism: hegemonic struggle and the incorporation of Bolivia and Indonesia. Int J Comp Sociol 58(4):275–303
Keating M, Hooghe L, Tatham M (2015) Bypassing the nation-state? Regions and the EU policy process. In: Richardson J, Mazey S (eds) European Union. Power and policy-making. Routledge, Abingdon
Kinnear S, Ogden I (2014) Planning the innovation agenda for sustainable development in resource regions: a central Queensland case study. Resour Policy 39:42–53
Kivinen S, Kotilainen J, Kumpula T (2020) Mining conflicts in the European Union: environmental and political perspectives. Fennia—Int J Geogr 198(1):163–179
Koch N, Perreault T (2019) Resource nationalism. Prog Hum Geogr 43(4):611–631
Kotilainen J (2018) Resilience of resource communities: perspectives and challenges. In: Marsden T (ed) The sage handbook of nature, vol. 1. Sage Publications, London
Kotilainen J (2021) Resource extraction, space and resilience. International perspectives. Routledge, London & New York
Kotilainen J, Eisto I, Vatanen E (2015) Uncovering mechanisms for resilience. Strategies to counter shrinkage in a peripheral city in Finland. Europ Plann Stud 23(1):53–68
Nem Singh TJ (2014) Towards post-neoliberal resource politics? The International Political Economy (IPE) of oil and copper in Brazil and Chile. New Polit Econ 19(3):329–358
Mazey S, Richardson J (2015) Shooting where the ducks are: EU lobbying and institutionalized promiscuity. In: Richardson J, Mazey S (eds) European Union. Power and policy-making. Routledge, Abingdon
Mireu (2021) Mining and metallurgy regions of the EU, https://mireu.eu/. Accessed 16 Feb 2021
Radkau J (2008) Nature and power: a global history of the environment. Cambridge University Press, New York
Rehner J, Baeza SA, Barton JR (2014) Chile's resource-based export boom and its outcomes: regional specialization, export stability and economic growth. Geoforum 56:35–45
Remix (2021) Smart and Green Mining Regions of EU https://www.interregeurope.eu/REMIX/. Accessed 16 February 2021.
Rittberger B, Winzen T (2015) The EU's multilevel parliamentary system. In: Richardson J, Mazey S (eds) European Union. Power and policy-making. Routledge, Abingdon
Schaffartzik A, Mayer A, Eisenmenger N, Krausmann F (2016) Global patterns of metal extractivism, 1950–2010: providing the bones for the industrial society's skeleton. Ecol Econ 122:101–110
Smith A (2014) The state, institutional frameworks and the dynamics of capital in global production networks. Prog Hum Geogr 39(3):290–315
Thomson R (2015) The distribution of power among the institutions. In: Richardson J, Mazey S (eds) European Union. Power and policy-making. Routledge, Abingdon
Tiess G (2011) General and international mineral policy. Focus: Europe. Springer, Vienna
Tiess G (2010) Minerals policy in Europe: some recent developments. Resour Policy 35(3):190–198

Veltmeyer H, Petras J (2014) The new extractivism. A post-neoliberal development model or imperialism of the twenty-first century? Zed Bookds, London

Wilson JD (2015) Understanding resource nationalism: economic dynamics and political institutions. Contemp Polit 21(4):399–416

Wonka A (2015) The European Commission. In: Richardson J, Mazey S (eds) European Union. Power and policy-making. Routledge, Abingdon

Juha Kotilainen is a human geographer and environmental policy University Lecturer at the University of Eastern Finland. He has published his research on environmental politics, resource exploitation debates, and local and regional development in, for example, the journals Land Use Policy, Forest Policy and Economics, The Extractive Industries and Society, Resources Policy, and European Planning Studies, and he is the author of the book Resource Extraction, Space and Resilience published by Routledge.

[illegible] (201[illegible]) [illegible] London (uncertain)
[illegible] (20[illegible]) Understanding [illegible] (uncertain) [illegible]
[illegible]
[illegible] (20[illegible]) [illegible]
[illegible]

[illegible]

Chapter 7
From the 'Pampas' to China: Scale and Space in the South American Soybean Complex

Maria Eugenia Giraudo

Abstract Soybean production in South America has expanded rapidly since the late 1990s, now covering an area equating approximately the size of Spain. This crop has expanded throughout the region, both by displacing other crops, and by incorporating new territories via transformation of ecosystems. The expansion of this production complex has seen the emergence of different scales of accumulation, distribution, organisation, and conflict: from local and rural communities undergoing socio-economic transformations, to the role of national states in governing the soy complex, to transnational capital investment, international price setting mechanisms, and global drivers for demand. Scale provides a helpful concept to understand the relationships of hierarchy between places and spaces, and their organisation around principles of capital accumulation. The multiscalarity of the soybean complex provides an example of how the imperatives of capitalist production at the global level orchestrate the geographical distribution of economic activity, and the emergence of resource peripheries, including soybean production. This chapter will first explore the theoretical elements of the concept of scale, followed by an examination of the multi-scalar process of expansion and consolidation of South America as a resource periphery, through the emergence of the soybean complex. The chapter will look at the tensions between national and global scales that led to these changes, as well as the impacts on regional dynamics, bodies and environment of the soybean production complex.

Keywords Scale · Commodities · Natural resources · Development · Space · South America

7.1 Introduction

The Southern Cone in South America has transformed, in the last 20 years, into the heart of soybean production globally. In that period, the oilseed has become

M. E. Giraudo (✉)
School of Government and International Affairs, Durham University, Durham, UK
e-mail: maria.e.giraudo@durham.ac.uk

F. Irarrazaval and M. Arias-Loyola (eds.), *Resource Peripheries in the Global Economy*, Economic Geography, https://doi.org/10.1007/978-3-030-84606-0_7

a global commodity and a significant driver of geographic, environmental, social, and economic transformations in the region. The extent of these transformations is dramatic. This was demonstrated by the devastating fires in the Amazon rainforest in 2019—a phenomenon that has been directly connected to the expansion of beef and soybean production in Brazil (Lai et al. 2019). The growth of soybean in South America is such that the oilseed now covers an area the size of Spain. Moreover, this trend is likely to continue, with Brazil expected to become the world's largest soybean producer by 2029 (OECD/FAO 2020, p. 139). Soybean has expanded through the landscape of South American countries—particularly in Brazil, Argentina, and Paraguay—taking over territories where other activities and crops had been produced before, and pushing the boundaries of the agricultural frontier through deforestation, road-building, dredging, and many other ecosystem-changing techniques.

The expansion of this production complex has seen the emergence of different scales of accumulation, distribution, organisation, and conflict: from the local transformations of rural communities, to the role of national states in governing the soy complex, to transnational capital investment, international price setting mechanisms, and global drivers for demand. The concept of scale serves as a tool for conceptualising places and the relationships between them around an organising principle (Sokol 2011). At the global level, it is the imperatives of capitalist production that orchestrate the geographical distribution of economic activity, including soybean production, and as such creates scalar hierarchies—determining, for example, the emergence of certain areas as resource peripheries. In areas of production—peripheral scales—this has meant a radical change in the systems and places of production, the availability of transport and storage networks, and in the ecological and geographical characteristics of these spaces.

This chapter will first explore the theoretical elements of the concept of scale and the scalar fix (Brenner 1998), followed by an examination of the soybean complex's multiple scalarity, and the tensions between the national, regional, and global scales over which will serve as the 'anchor point' of capital accumulation. The second section will look at the emergence of the soybean complex and the multi-scalar process it entailed: namely, how the conflagration of dynamics at the national and global level, and the changes in hierarchies of the global economy, created the conditions for the transformation of spaces of production in South America, both within, and as global peripheries for primary production. The final section will explore the consolidation of the soybean complex in Argentina, Brazil, and Paraguay, and how this has led to the emergence of a new regional scale of soybean production, one with significant environmental and social impacts.

7.2 Space, Scale, and the Political Economy of Soybean Production

The expansion of soybean needs to be understood as a spatial phenomenon, specifically in light of the changes witnessed in the past two decades. Since the mid-1990s, soybean production in South America has expanded rapidly, displacing other productive sectors and changing the landscapes of affected regions. The total area planted with soybean almost tripled in Brazil and Paraguay between 2000 and 2019 and doubled in Argentina in the same period, an expansion that is the result of both the replacement of other crops and the incorporation of new land into the agricultural frontier (FAOSTAT 2021). This growth is not only reflected in production volumes, but it has been expressed by fundamental physical changes in the geography of the region, such as expanding deforestation, the construction of new roads, river dredging, and the opening of new ports. These physical modifications associated with the growth of the soybean industry—in other words, its consolidation as a resource periphery—have left an indelible imprint on the region's landscape. Furthermore, various financial instruments and market infrastructures have been developed, creating extensive virtual connections between the Southern Cone and economic centres around the globe. These intangible innovations, associated with the globalisation of the soybean trade, have effectively shrunk the relative space separating Latin America from other world regions. All of this suggests that the soybean boom must be grasped as an inherently *spatial* phenomenon (Fig. 7.1).

To understand the extent and nature of the spatial transformations created by the soybean complex, and the multi-scalar feature of the phenomenon, we need to study it as a process of capital accumulation. Capital accumulation, and the political regulation of this process, develops in particular 'spatio-temporal matrices' (Jessop 2000, p. 327), leading to a spatial structuring of social life (Soja 1989, p. 127). With each new wave of capitalist expansion, various patterns of spatial organisation arise that enable the creation and distribution of profit. As Neil Brenner explains:

Each successive round of capitalist industrialization has therefore been premised upon socially produced geographical infrastructures that enable the accelerated

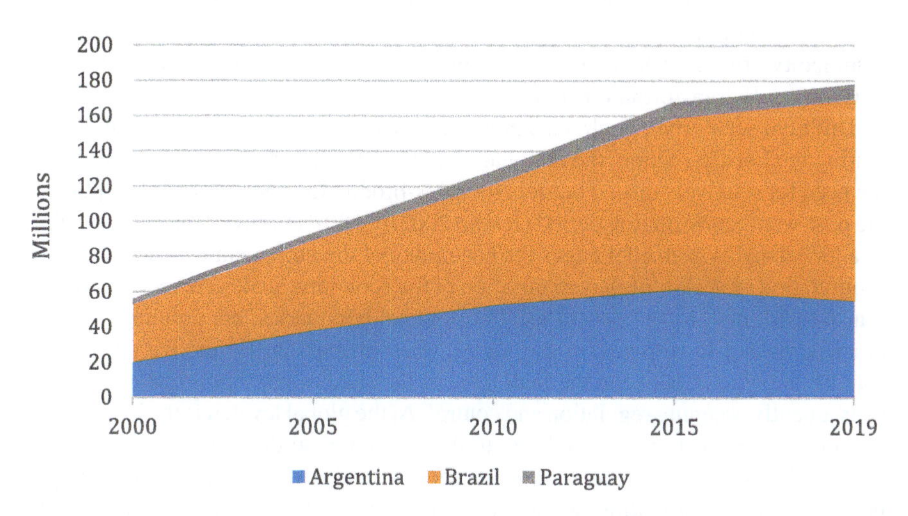

Fig. 7.1 Soybean production in tonnes, 2000–2019. *Source* FAOSTAT (2021)

circulation of capital through global space. In this sense, as Harvey notes, 'spatial organisation is necessary to overcome space' (Harvey, p. 145). (Brenner 1999, *p. 43).*

This spatial organisation also entails a re-organisation of the different scales at which action takes place, as well as the relationship between scales and the role they play in the process of capital accumulation. And within this, resource peripheries play a key role in enabling capital accumulation in the core. As an analytical framework, the concept of 'scale' was introduced with the rise of globalisation studies, as the transformative effects of 'rescaling' began to be felt (Herod 2011, p. 1). As a tool that allows for the spatial conceptualisation of different types of social activity, the notion of scale entails a central organising principle. For example, scalar organisation in the military—brigade, battalion, company, etc.—is a function of security requirements. In the case of the global political economy, the principle of scalar organisation is inherently linked with the accumulation imperatives of capitalism (Smith 2010).

Scale is usually understood as a 'container'—the location where social activities occur and within which the production of space takes place (Brenner 1998, p. 460). The urban, the national, and the global, appear as 'ontologically given' arenas. However, as Smith points out, there is nothing natural about these various scales—they have no prior existence outside of the historically specific social relations that give them meaning (Smith 1992, p. 73). The author himself recognises that to only identify these three scales was limited, and that more attention should be paid to others, such as the household (Smith 2011, p. 261). In this sense, the understanding of space as socially produced forces us to conceptualise scale, 'the most elemental differentiation of geographical space', as established and produced precisely through social activity (Smith 1992, p. 73). The way in which we define an event is determined by the manner in which social, economic, and political life is organised. For example, a drought in the US may be considered a local issue with regards to its immediate impact on farmers, a national issue due to its effects on US agricultural exports, or a global issue because of its impact on international commodity prices. Scale, then, is an essentially *social* spatial differentiation, and is thus subject to transformation as social relations develop.

Different scales then might take prevalence in different regimes of accumulation. During the Fordist regime, the national scale was the dominant one, namely, due to the need for relatively closed borders for mass-production and mass-consumption in the post-war reconstruction period (Jessop 2003, p. 180). However, the consolidation of a globalised system challenges the hegemony of the nation-state, as we witness a proliferation of scales for the organisation of the economy. Scales are redesigned and transformed, not on the basis of legal or political boundaries, but around economic efficiencies and in response to the logics of profit making. Extractive peripheries can then emerge in between the fault lines of the state or other political authorities, consequently escaping regulation and control. At the global level, it is the imperatives of capitalist production that orchestrate the geographical distribution of economic activity. In this context, the emergence of new scales is an indicator of a larger process of *rescaling* within the global political economy, whereby new spaces and temporalities for the organisation of capital accumulation emerge and challenge the primacy of those already established. This has been framed in different terms by various authors, from the 'politics of scale' (Brenner 2001), to the 'production of scale' (Smith 2010), to the 'relativisation of scale' (Jessop 2003). Despite their differences, all of these approaches advance the notion of a competitive struggle between existing and emerging scales over which will become the 'new anchor point around which other scales can be organized' (Jessop 2003, p. 181). The proliferation of new scales reflects an attempt to restructure the spatial roots of politico-economic, as well as social and cultural, processes. Such rescaling has taken place in the spheres of both capital and labour: from the multinationalisation of corporations to the global expansion of labour movements, such as the Global Union Federation (GUF) or anti-sweatshop campaigns (Herod 2011, p. 22).

The configuration of scales that emerges from the dynamics of capital accumulation, then, is a 'historically specific, multi-tiered territorial-organization' (Brenner 1998, p. 464). Brenner highlights the scalar dimension of these spatial forms, or what he calls a 'scalar fix' (Brenner 1998), building on Harvey's conceptualisation of 'spatial fix', which denotes the tendency of capital to search for different coherent and stable geographical configurations in the aftermath of overaccumulation crises. The continued reproduction of capital after a crisis is achieved through the emergence of a new bundle of hierarchically organised boundaries, or 'scalar fixes', which can facilitate renewed capital accumulation (Brenner 1998, p. 464). While Harvey's concept of spatial fix presupposes the relocation of economic activity, Brenner adds that this relocation also implies a different hierarchy of scales, as for example, the national scale may be superseded by supranational or global arrangements, or the urban scale might emerge as the main locus of accumulation.

The new constellation of scalar arrangements that emerge following economic crises should be understood as scalar fixes, which allow capital accumulation to continue in a spatially reconfigured manner. But scale should not be studied in isolation or as the only relevant dimension of capitalist phenomena. Rather, as Jessop et al. (2008) point out, it exists in an entanglement of territories, places, and networks. The authors propose a strategic-relational approach to territories, places, scales, and networks (TPSN) as a lens through which to highlight the way in which this entanglement of sociospatial relations both reflect and attempt to solve the contradictions inherent to capital accumulation (2008). In terms of this chapter's case study, the spatial composition of the soybean complex attempts to solve the recurrent crises of capitalism by creating a new 'spatial fix', which at the same time, reproduces the contradictions that lead capitalism to its crisis.

The expansion of the soybean production complex in South America has seen the emergence of different scales of accumulation, distribution, organisation, and conflict. We can see how port facilities serving national production, like the Gran Rosario area in Argentina, now become key regional logistic hubs; or how fluidity in ownership and land access in border areas, such as that between Brazil and Paraguay, transform these in spaces of contestation. The different aspects of this complex embody a multi-scalar dimension, as well as processes of rescaling. Cross-border transactions often escape the control of national governments, and the ownership of capital is not determined by national borders, as illustrated by statements of many officials and producers in Brazil: 'soy planted in Paraguay is Brazilian' (interview

APROSOJA, 2014). Similarly, transport infrastructure projects are determined by the demands of the global market, rather than the principles of the national economy or the needs of local populations. A great proportion of newly built roads in the Southern Cone are designed for commodity transport, with the movement of national citizens considered incidental. As Jessop et al. (2008) pointed out, in the sociospatial configuration of the soybean complex, multiple scales are 'entangled' with networks, places, territories, and even new ecosystems that are created in the process of capital accumulation. This emergent scalar configuration transforms places of production into sites of global extraction in benefit of the core, and their role and characteristic as space is determined by their relationship to other scales—as providers of raw materials, as enablers of profits, as well as spaces of marginalisation and contestation. The next section will explore the emergence of the soybean complex and the relationship between the national and new scales of capital organisation it involved.

7.3 The Soybean Complex in South America

Soybean production in South America has expanded rapidly since the late 1990s. Unlike traditional extractive activities such as mining or hydrocarbon exploitation, soybean production and industrial agriculture in general expands throughout space, incorporating new territories, new land, as it advances. The expansion of this production complex has seen the emergence of different scales and places of accumulation, distribution, organisation, and conflict: local, national, and global dynamics are entangled in the creation and consolidation of this particular productive and accumulation system.

In that sense, the soybean complex encompasses all the dimensions that Jessop et al. (2008) discuss: territory, place, scale, networks. It is, then a 'polymorphous' phenomenon (Jessop et al. 2008), one that constitutes a 'spatio-temporal fix' in the process of capital accumulation. In addition, the temporal dimension requires attention, as the soybean complex is also traversed by multiple temporalities: 'the annihilation of space by time' that Marx (1973) discussed indicates the need for fast turnover times to shrink the distances between resource peripheries, zones of industrial production, and consumer markets. But there is also another temporality: the 'slow violence' (Nixon 2011) against bodies and nature that the new reconfiguration of space, nature, and relations that the soybean complex has exerted.

7.3.1 Soybean as a Global Commodity

These changes in productive dynamics at the national level developed in conjunction with transformations in the global economy, namely, the emergence of a commodity boom in the early 2000s that changed terms of trade for food and natural resource exporters. An increase in demand pushed up commodity prices in 2001–2002,

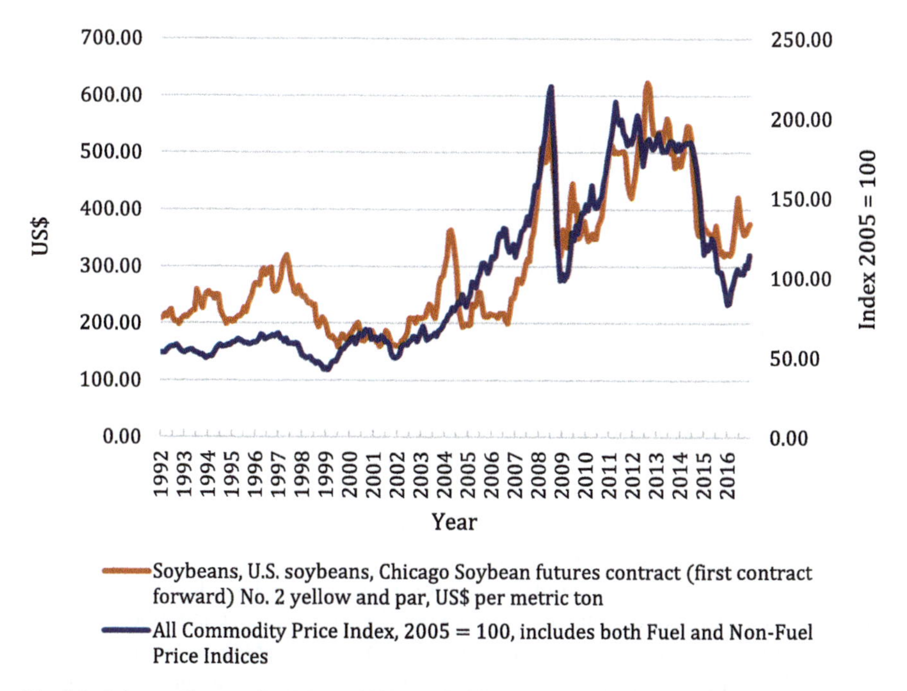

Fig. 7.2 Primary Commodity Prices, 1992–2017. 2017 *Source* IMF Commodity Prices,

reaching a peak in 2007–2008. During this commodity boom, soybean emerged as one of the most important assets, as its price rose even higher than wheat and maize—the other cereals that experienced significant price peaks in this period (IMF Commodity Prices 2017). As shown in Fig. 7.2, between 2004 and 2008, the price of soybean increased by almost 65%, following the movements of the All Commodity Price Index. In this context, soybean emerged globally as one of the fastest growing sectors in agricultural trade, with export values reaching almost US$ 100 billion in 2013 (FAOSTAT 2021). As a rapidly expanding crop with volatile and relatively high international prices, soybean became a very attractive investment for global capital.

Between 1990 and 2014, there was a three-fold increase in the global production of soybean, which translated into a growth in value from US$ 26 billion[1] to over US$ 67.5 billion (FAOSTAT 2021). Global production is dominated by the US, Brazil, and Argentina, which produce more than 75% of the world's output, with Brazil and the U.S. expected to account for two-thirds of global soybean production by 2029 (OECD/FAO 2020, p. 139). India, Paraguay, and China follow in production volumes, resulting in a concentration of 90% of global production in these six countries. According to the latest OECD-FAO report (2020), soybean grew at an average of 4.0% per year in the last 10 years (2010–2020). Soybean has now become a key agricultural commodity, with similar trade and production levels to those of historical food staples such as maize and wheat. Not only has soybean experienced a huge growth in the

[1] Gross Production Value in constant 2004–2006 million US$.

last 20 years, but production is still expected to rise at an annual average of 1.3% between 2020 and 2029, with expansion of harvested area accounting for a third of this increase (OECF/FAO 2020, p. 139).

The rise of soybean as a global commodity cannot be explained without looking at the drivers for demand. The main soy importers are China and the European Union (EU). In 2014, China imported 69 million tonnes of raw soybean and 1.4 million tonnes in soybean oil, while the EU received over 12 million tonnes of soybean and 19 million tonnes of soybean meal in the same period (Foreign Agricultural Service/USDA 2017). For that year, the total value of soybean trade—including oil, meal, and beans—was US$ 109 billion (Chatham House 2017). In particular, China has played a key role in pushing the global demand for soybean upward, driven by improvements in diets of its population, combined with a food security strategy that emphasised domestic production of other grains, eventually creating a 'soybean nexus' with South American countries (Giraudo 2020). As Jepson (2020, p. 79) highlights, the impact of Chinese demand of soybean has been such on global markets that the oilseed and the impacts of its global boom can be equated with that of non-agricultural natural resource exports, such as copper.

The soybean boom and the transformations it has created in the landscapes, livelihoods, systems of production, and national development paths, cannot be explained without understanding the complex interaction of processes at the national and global level, with changing hierarchies in the scalar organisation of the economy, as capital searched for a new 'scalar fix'. The next section will unpack how the consolidation of the soybean complex as a spatio-temporal fix for capital accumulation created new scales, produced nature, and transformed bodies.

7.3.2 National Territories and the Expansion of Soy in South America

Unlike other natural resources, the mobility of soybean (originally an Asian crop) and its material conditions (which allowed for its high adaptability) provided a transnational character to its production. The oilseed began to expand rapidly across the Southern Cone in the 1990s and eventually became a key crop in the rural sector, transforming landscapes, territories, and livelihoods as it advanced. The emergence of the soybean complex in South America is the result of multi-scalar processes that include shifting hierarchies at the global level, governance transformations at the national scale, and changing relations of production at the local level.

Locally, for farmers, soybean became an easy and cost-effective crop. Across the borders of several South American countries, the 'technological package' of soybean was adopted: genetically modified 'Round-up Ready' (RR) soybean, agrochemicals (particularly the weed-control pesticide glyphosate), and direct seeding or 'no-tillage' as a production technique (Turzi 2011; Leguizamon 2016; Phélinas and Choumert 2017). This particular package provided very high yields, low costs, and as a result, high profitability for a crop with multiple uses. This combination of factors and namely, the huge potential for profitability were key material processes in the multiplication of sites of soybean production, and as such, the consolidation of this area as a resource periphery. According to Mariano Turzi (2011, p. 61), the adoption of GM soybean seeds 'spread like bushfire', and through displacement of other crops—like cotton in Paraguay, or maize and wheat in Argentina—transformed the features of agricultural production in these countries. As soybean advanced over other crops and activities, the region increasingly lost economic diversity, developing a growing dependence on the oilseed, and hence a developmental 'straitjacket' found in resource peripheries (Barton et al. 2008, p. 26).

However, processes at the national level set the conditions for these transformations, and many of the changes in local production have specific characteristics in each of the national territories. Argentina was the first country in the region to legally approve the use of Genetically Modified (GM) seeds. This incorporation and the productive model that accompanied it was enabled by an institutional context of deregulation—legitimised in a global context that promoted a neoliberal model of development (Leguizamón 2014). This framework also allowed the increase of short-term farming through leasing, which in turn encouraged large-scale farming while the property structure of the country maintained the existence of medium-size farms. Finally, the high profitability of soybean not only incentivised intensive farming and mono-cropping in the region of the *pampas*, the traditional farming, and cattle area, but also promoted the expansion of the soybean frontier towards the North-East and North-West through deforestation, allowing for a greater concentration of land ownership within agriculture.

In Brazil, soybean was introduced by Japanese migrants who would plant it for their own consumption during the early twentieth century (Schlesinger 2013; Oliveira and Schneider 2016). But production soared towards the end of the 1990s, mainly through the significant movement of farmers from the Southern states—such as Rio Grande do Sul—towards the Centre and Centre-West of the country, in the biome known as *Cerrado*, in search of larger properties. Key in the expansion of soybean into different areas of Brazil was the role of the state agency for agricultural research, EMBRAPA (Empresa Brasileira de Pesquisa Agropecuária), which developed soybean varieties that could adapt to the different climates, hence creating new frontiers for soybean production (interview APROSOJA, 2014; Cattelan 2012). Similar to Argentina, the emergence of the soybean complex would not have been possible without state intervention.

As for Paraguay, the expansion of soybean production has gone hand in hand with the increasing presence of foreign farmers, chiefly Brazilian but also German and Japanese settlers, during the 1970s (Ortega 2015). The most drastic growth of the oilseed, however, came in the 1990s, triggered by two factors: the permeability of the border that allowed the unlicensed entrance of GM soy coming from Argentina; and the influx of investors from Brazil in search of cheaper lands, motivated by the modernisation of Brazilian agriculture (Ezquerro-Cañete 2016; Ortega 2015). Throughout this process, the absence of regulation from the Paraguayan state marked the development of the soybean sector and the relationship between both (Hetherington 2020).

Rather than being a passive actor, the state has played a key role in creating the conditions that incentivised the advancement of the soybean complex and that contributed to the permeability of capital across the borders of Argentina, Brazil, and Paraguay—the region's main soybean producers. This challenges views that consider the state as 'retreating', particularly during the 1990s, when capitalism's restructuring led to new scalar hierarchies, with a prevalence of the global (Jessop et al. 2008, p. 390). Instead, the state, and thus the national scale, was key in the process of rescaling required by capitalist transformations.

7.4 Consolidating the Soybean Complex: Production of Scale, Nature, and Bodies

The previous section explored how the process of accumulation linked to the soybean complex is one that emerges from a rich scalar dynamic between spaces of production, national economies, and the changing dynamics of the global economy. Underlying these processes is an understanding that space—and territories, places, scales, and networks—is not a static element, but it is *relational,* and as such, the result of social relations, in particular *capitalist* social relations (Brenner 1999, p. 43). Smith (2010) focused on how the contradictions of capitalism created patterns of uneven development across the globe, as capital floods out of certain zones and into others in search of profit. It is in this quest for higher profitability that capitalism produces space—new spaces for production and for the accumulation of capital. Not only that, but nature is also subject to the transformations of these new spaces (Smith 2010, p. 50). As accumulation through soybean production advances, we see new 'natures' being produced: through deforestation, fires, road-building, dredging of rivers, and the disregard for national borders. The consolidation of the soybean complex has created new scales of production, destroyed and created new ecosystems, and even transformed bodies.

7.4.1 Reconfiguring the Regional Scale: Redesigning Landscapes and the Logistics Turn

The interaction between dynamics at the national and global scale transformed existing agricultural landscapes and created new ones in the expansion of soybean production in the Southern Cone. As previously mentioned, the 'soyisation' of Argentina, Brazil, and Paraguay occurred through a shift from other crops into soybean, as well as the incorporation of new lands into the agricultural frontier.

In Paraguay alone, agricultural land increased by 67% between 1980 and 2018 (CEPALSTAT 2021).

While there were national dynamics at play, the soy model rapidly expanded throughout the borders of South American nations, particularly in Argentina, Brazil, and Paraguay, but increasingly so in Uruguay and Bolivia. The first component of this model to move quickly across national borders was the genetically modified RR soy produced by Monsanto. Approved by Argentina in 1996, RR soybean circulated and was used for production in Paraguay and Brazil well before these countries formally legalised its use, which only occurred between 2002 and 2005 (Hetherington 2013; Oliveira and Hecht 2016; Ezquerro-Cañete 2016). Until then, the GM variety would be transported via contraband from Argentina in 'white bags'. The circulation of RR soybean was then the first step towards the adoption, at a transnational level, of soybean's 'technological package'. This package, consisting of GM soybean seeds, agrochemicals, and no-till sowing, was key in the expansion of soybean monoculture, and transformed the 'ecological, economic and political parameters of agricultural production' (Turzi 2011).

Turzi (2011) and other authors have understood the networks and integrated chain of soybean, as well as the hegemony of this productive complex in the Southern Cone as the emergence of a new, *transnational* geopolitical and economic entity: the 'Soybean Republic' or 'Soylandia' (Rulli 2007; Pearce 2012). This reflects a vision of national borders in flux, to the point that the state seems to lose control of the social and economic activities occurring within its territory. These studies do, like much of scholarship in International Political Economy, recognise the existence of spatiality, either implicitly by highlighting the emergence of this transnational entity and discussing *national* strategies, or occasionally explicitly by making reference to a *territory* with certain characteristics (cf. Macartney and Shields 2011). In other words, the soybean complex as an accumulation regime has given rise to a new regional scale organised around the production, processing, transportation, and trade of soybean from the resource peripheries in South America to global markets.

Several mechanisms can be identified through which the space in the Southern Cone of the country was redesigned around soybean and re-scaled as a regional, trans-border complex. The first, which we have already discussed, is the changes in the landscape that the expansion of soybean production has created, both by expelling other crops and pastoral activities and by incorporating new lands. Deregulation in agricultural sector and state promotion of technological innovation made soybean a very attractive productive package, with low costs and high yield as well as great capacity for adaptation to different climates, allowing it to quickly expand across traditional agriculturally productive areas. More importantly, consistent high international prices—particularly during the commodity boom—and high profitability have pushed the agricultural frontier into new areas, through deforestation (either by logging or fires), increasingly into border areas. As Fig. 7.3 shows, the extension of soybean production has created a new map of South America, and transformed the region into a homogenous landscape of soybean monoculture.

A second mechanism has been the development of a network of transport infrastructure that enables the movement of this commodity towards its export markets.

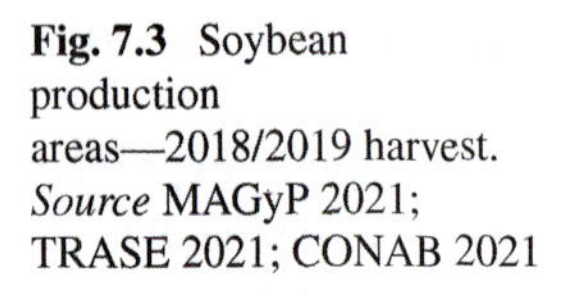
Fig. 7.3 Soybean production areas—2018/2019 harvest. *Source* MAGyP 2021; TRASE 2021; CONAB 2021

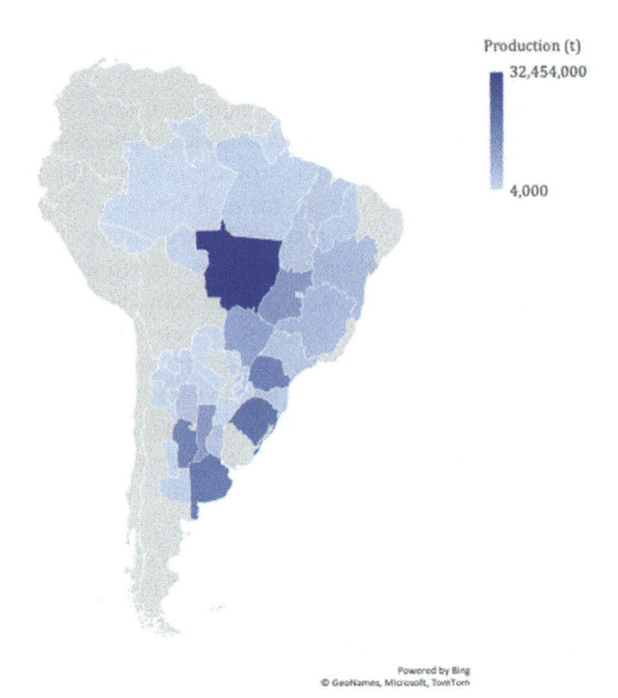

The accumulation of capital in the soybean complex cannot be realised if the product is not moved towards its final consumer destinations. On average, 80% of soybean produced in Argentina, Brazil, and Paraguay is destined for international markets (FAOSTAT 2021). With this in mind, the significance of infrastructure and the wider built environment for the viability of the soybean complex cannot be overstated. The network of roads, railroads, waterways, and ports that extend throughout the region have the aim of 'annihilating space by time' (Marx 1973), that is, increasing turnover time in order to reduce costs and expand profit margins. The cases of the Gran Rosario logistics complex in Argentina, the BR-163 road in Brazil, and the Paraguay-Paraná river way are key examples of how the soybean economy has created a particular built environment—rooted in specific national territories—for the sole purpose of faster circulation of larger volumes of soybean from its production sites in the Southern Cone to different points in the global value chain.

Finally, as the soybean production model expanded and consolidated as the dominant form of agricultural production, transnational corporate actors, particularly agrochemical companies, were 'empowered' by the success of this technological package (Turzi 2011, p. 62). Through the development of innovations in biotechnology and agrochemicals, and enabled by intellectual property rights, transnational companies like Monsanto and Syngenta became key actors in the region. There is a territorialisation of their presence through the establishment of subsidiaries, the creation of networks with local farmers, producers' associations, and governments, among others, and even by contributing to the narrative of a new scale of economic organisation and production articulated around soybean and its technological package. A clear example of this is the advertisement published in 2003 in some of the biggest newspapers of Argentina by the agrochemical company Syngenta, in which it announced that 'soybean knows no borders', as a map showed the flowing borders of the 'United Republics of Soybean' across the territories of Argentina, Brazil, Paraguay, Uruguay, and Bolivia (Grain 2013).

Agrochemical companies are not the only corporate actors that have contributed to the reconfiguration of space that constitutes the soybean complex. Much of the profits that this complex generates are created through trading the oilseed in bulk. Agricultural trading companies, mainly the so called 'ABCDs'—an acronym for the four biggest grain traders, Archer Daniels Midland, Bunge, Cargill, and Louis Dreyfus—feature as the main exporters of soybean in all aforementioned countries. To do so, it is key for the development of processing and transport infrastructure that can deterritorialise soybean production—that is, make it realise its purpose in export markets.

7.4.2 *Bodies and Nature: the Environmental and Social Impacts of the Soybean Complex*

In his book *Uneven Development*, geographer Neil Smith explains how, in its quest for universality and profitability, capitalism creates its own limitations:

*It creates a scarcity of needed resources, impoverishes the quality of those resources not yet devoured, breeds new diseases, (…) pollutes the entire environment that we must consume in order to reproduce, and in the daily work process threatens the very existence of those who produce the vital social wealth (*Smith 2010, *p. 84).*

The economic and geographic transformation that the expansion and consolidation of the soybean complex has generated cannot ignore the impact it has pressed on nature and bodies. The 'production of nature' (Smith 2010) in the soybean complex is also a scalar phenomenon, with impacts that range from one of the largest and most important ecosystems in the planet, the Amazon basin, to remote rural areas in Córdoba, Argentina. In the same way that the entanglement of territories, places, scales, and networks creates the conditions for capital accumulation, while attempting to solve capitalism's contradictions (Jessop et al. 2008), the production (or destruction) of nature embodies these contradictions, as it is both necessary for the expansion of capitalism, and creates its own demise (Smith 2010). While examples are numerous, two cases will help illustrate how these contradictions materialise in nature and bodies.

While initially incorporated to add nitrogen to the soil, and as such to help its nutritional composition, the shift in agricultural production towards monoculture, and thus the elimination of crop rotation that had ensured the health of the underground, began to create serious erosion problems. Moreover, the expansion of soybean into newly incorporated agricultural land, which implied ecosystem transformations, eventually started to generate very visible reactions from the environment. One of these reactions was the appearance of a network of rivers in San Luis in central Argentina. In 2015, local residents noticed a ravine that had emerged in their land, one that eventually grew into a 25 km long river (Goñi 2018). Experts at the Universidad de San Luis attributed this to a combination of factors, including the quality of the soil, and the fact that intensive agriculture, namely, maize and soybeans had replaced land which had always been covered by forests and grasslands. The 'production of nature' in the soybean complex, in this case, takes on a very literal meaning.

These transformations in the environment have serious implications for those who inhabit it—both human and non-human lives. In the case of the 'New River' in San Luis, the instability of the climate and the erosion of the soil have had terrible impacts on the livelihoods of farmers in the area, who might see their farm lands suddenly flooded and covered by sediment from the river (Goñi 2018). Even more striking has been the impact that soybean's technological package has had on the bodies of the people living surrounded by seas of soybean fields. In Paraguay, a professor of clinical diagnosis, Dr. Insfrán, circulated a report that highlighted the dramatic increase in blood cancer in Paraguay, particularly in the countryside, suggesting a link between these anomalies and the use of GM crops and agrochemicals (Hetherington 2020). In Córdoba, Argentina, a group of mothers in the peri-urban town of Ituzaingó decided to map a number of health problems that had multiplied in the town, which was surrounded by fields of soybean and where the indiscriminate use of agrochemicals meant that houses and roads would end up covered in pesticides. These disorders included cancer, malformations, and respiratory diseases (Berger and Ortega 2010). While further studies are required to explore more accurately the connection between agrochemicals used in soybean production and these numerous diseases, a task that has proven extremely difficult given the power of the soybean industry (Hetherington 2020), across the region there is increasing evidence of the link between soybean's technological package and the suffering of entire population living on the margins of the soy model.

7.5 Conclusion

In just 10 years—between 2000 and 2010—the planted area of soybean increased by over 40% in Argentina and Brazil, and a remarkable 95% in Paraguay. As previously mentioned, this now equates to an area about the size of Spain, entirely dedicated to the soybean mono-cropping, meaning the continuous production of the same crop on the same land, without rotation.[2] Expansion in the planted surface and increased

---

[2] Calculation based on conversion of hectares into square kilometres. While soybean production occupies 50 million hectares, which translates into 500,000 sq. km., the total surface area of Spain is 505,992 sq. km.

yield contributed to a 100% growth in production in that same decade. This rapid increase in production was possible by both expanding the arable land in the region, and by converting land from other crops into soybean (Giancola et al. 2009, p. 32).

The proliferation of this crop across the borders of these countries was the expression of a process of homogenisation throughout the Southern Cone. The expansion of soybean embodied the emergence of new relations of production that transformed places and landscapes, attached to regional and global networks, while also being situated in national territories and shaped by historical territorial processes. This reconfiguration reinforces and deepens the positioning of the region as a resource periphery and embodies the transformation that the productive spaces endure. The soybean complex is a polymorphous entity that constitutes a new spatio-temporal matrix of capitalism (Jessop et al. 2008). The compound of places, territories, networks, and scales that make up the soybean complex in South America are the expression of a new spatiality of capitalism in the region. That is, the geography of the subcontinent is reshaped and restructured as a result of capitalist imperatives, which have found in soybean a source of profits. As capital rushed into agricultural production in South America, space—territories, places, networks, and scales—has been reorganised in line with profitability dictates. The emerging spatiality also carries a new temporality, one driven by the 'annihilation of space by time' (Marx 1973, p. 539), as reducing the distances between places of production and consumption is achieved by ensuring the fast and efficient circulation of commodities.

This chapter has demonstrated the rich, multi-scalar nature of the soybean complex in South America, from the transformation of local practices of production, through the development of networks and production chain integration at a regional scale, to identifying the dynamics at the national and global level driving these processes. We have explored the multifaceted and complex ways in which the economic, political, social, and environmental landscape of the region has been transformed in a way that reproduces and consolidates the position of South America as a provider of natural resources in the global political economy.

References

Barton JR, Gwynne RN, Murray WE (2008) Transformations in resource peripheries: an analysis of the chilean experience. Area 40:24–33

Berger M, Ortega F (2010) Poblaciones expuestas a agrotóxicos: autoorganizacion ciudadana en la defensa de la vida y la salud, Ciudad de Córdoba Argentina. Physis Revista De Saúde Coletiva 20(1):119–143

Brenner N (1998) Between fixity and motion: accumulation, territorial organization and the historical geography of spatial scales. Environ Plann d: Soc Space 16(4):459–481. https://doi.org/10.1068/d160459

Brenner N (1999) Beyond state-centrism? Space, territoriality, and geographical scale in globalization studies. Theory Soc 28(1):39–78

Brenner N (2001) The limits to scale? Methodological reflections on scalar structuration. Prog Hum Geogr 25(4):591–614. https://doi.org/10.1191/030913201682688959

Cattelan AJ (2012) Breeding programmes and availability of non-GM IP seeds for farmers in Brazil. Separatas, EMBRAPA Brasil
CEPALSTAT (2021) Databases and statistical publications, economic commission for latin America and the caribbean. Retrieved https://estadisticas.cepal.org/cepalstat/Portada.html?idioma=english
Chatham House (2017) Resourcetrade.Earth. Retrieved http://resourcetrade.earth/
CONAB (2021) Companhia Nacional de Abastecimento, Governo Federal de Brasil. Retrieved https://portaldeinformacoes.conab.gov.br/safra-serie-historica-graos.html
Ezquerro-Cañete A (2016) Poisoned, dispossessed and excluded: a critique of the neoliberal soy regime in Paraguay. J Agrar Chang 16(4):702–710. https://doi.org/10.1111/joac.12164
FAOSTAT (2021) FAOSTAT Statistics Database. Retrieved http://faostat3.fao.org/faostat-gateway/go/to/home/E
FONPLATA (2017) Cuenca Del Plata [the Plata Basin]
Foreign Agricultural Service / USDA (2017) Soybeans & oil crops overview. Retrieved https://www.ers.usda.gov/topics/crops/soybeans-oil-crops/
Giancola SI, Salvador ML, Covacevich M, Iturrioz G (2009) Analisis de La Cadena de Soja En La Argentina. Instituto Nacional de Tecnologia Agropecuaria
Giraudo ME (2020) Dependent development in South America: China and the Soybean Nexus. J Agrar Chang 20(1):60–78. https://doi.org/10.1111/joac.12333
Goñi U (2018) When nature says 'Enough!': The river that appeared overnight in Argentina. (2018). The guardian. http://www.theguardian.com/world/2018/apr/01/argentina-new-river-soya-beans
Grain (2013) La Republica de la Soja recargada. Against the grain. https://www.grain.org/article/entries/4739-la-republica-unida-de-la-soja-recargada
Herod A (2011) Scale. Routledge, Oxon
Hetherington K (2013) Beans before the law: knowledge practices, responsibility, and the paraguayan soy boom. Cult Anthropol 28(1):65–85. https://doi.org/10.1111/j.1548-1360.2012.01173.x
Hetherington K (2020) The government of beans: regulating life in the age of monocrops. Duke University Press, Durham
IMF (2017) IMF primary commodity prices. Retrieved 30 May 2016
Jepson N (2020). In China's wake: how the commodity boom transformed development strategies in the global South. Columbia University Press, New York
Jessop B (2000) The crisis of the national spatio-temporal fix and the tendential ecological dominance of globalizing capitalism. Int J Urban Reg Res 24(2):323–360. https://doi.org/10.1111/1468-2427.00251
Jessop B (2003) The political economy of scale and the construction of cross-border micro-regions. In: Theories of new regionalism. Springer, Berlin, pp 179–96
Jessop B, Brenner N, Jones M (2008) Theorizing sociospatial relations. Environ Plann d: Soc Space 26(3):389–401. https://doi.org/10.1068/d9107
Lai KKR, Lu D, Migliozzi B (2019) What satellite imagery tells us about the amazon rain forest fires. 24 August 2019. The New York Times
Leguizamón A (2014) Modifying argentina: GM soy and socio-environmental change. Geoforum 53:149–160. https://doi.org/10.1016/j.geoforum.2013.04.001
Leguizamón A (2016) Environmental injustice in argentina: struggles against genetically modified soy. J Agrar Chang 16(4):684–692. https://doi.org/10.1111/joac.12163
Macartney H, Shields S (2011) Finding space in critical IPE: a scalar-relational approach. J Int Relat Dev 14(3):375–383. https://doi.org/10.1057/jird.2011.9
MAGyP (2021) Ministerio de Agricultura, Ganadería y Pesca, Presidencia de la Nación Argentina. Retrieved https://www.magyp.gob.ar/sitio/areas/estimaciones/monitor/
Marx K (1973) Grundrisse : foundations of the critique of political economy. Harmondsworth: Penguin in association with New Left Review
Nixon R (2011) Slow violence and the environmentalism of the poor. Harvard University Press

OECD and Food and Agriculture Organization of the United Nations (2020) OECD-FAO Agricultural Outlook 2020–2029
Oliveira G, Hecht S (2016) Sacred groves, sacrifice zones and soy production: globalization, intensification and neo-nature in South America. J Peasant Stud 43:251–285. https://doi.org/10.1080/03066150.2016.1146705
Oliveira GDT, Schneider M (2016) The politics of flexing soybeans: China, Brazil and global agroindustrial restructuring. J Peasant Stud 43(1):167–194. https://doi.org/10.1080/03066150.2014.993625
Ortega G (2015) Incorporacion Del Paraguay al Mundo de Los Agronegocios. In: Palau M (ed) Con la soja al cuello. Informe sobre Agronegocios 2013–2015. BASE IS, Asuncion
Pearce F (2012). The Landgrabbers: the new fight over who owns the Earth. Random House
Phélinas P, Choumert J (2017) Is GM soybean cultivation in argentina sustainable? World Dev 99:452–462. https://doi.org/10.1016/j.worlddev.2017.05.033
Rulli J (ed) (2007) Repúblicas Unidas de La Soja. Realidades Sobre La Produccion de Soja En America Del Sur. Grupo de Reflexion Rural
Schlesinger S (2013) Dois Casos Sérios Em Mato Grosso. A Soja Em Lucas Do Rio Verde e a Cana-de-Açúcar Em Barra Do Bugres. Mato Grosso: FORMAD Forum Mato-grossense do Meio Ambiente e Desenvolvimento
Smith N (1992) Geography, difference and the politics of scale. In: Doherty J, Graham E, Malek M (eds) Postmodernism and the Social Sciences. London, Palgrave Macmillan
Smith N (2010) Uneven development: nature, capital, and the production of space, 3rd edn. Verso, London
Smith N (2011) Uneven development redux. New Pol Eco 16(2):261–265. https://doi.org/10.1080/13563467.2011.542804
Soja EW (1989) Postmodern geographies: the reassertion of space in critical social theory. Verso
Sokol M (2011) Economic geographies of globalisation: a short introduction. Edward Elgar, Cheltenham, UK, Northampton, MA
TRASE (2021) Transparency for sustainable economics. Retrieved https://trase.earth/
Turzi M (2011) The soybean republic spotlight on resources. Yale J Int Aff 6(2):59–68

Maria Eugenia Giraudo is Assistant Professor in International Political Economy at the School of Government and International Affairs at Durham University. She obtained her PhD from the University of Warwick. Her research addresses the uneven outcomes of the latest commodity boom and accompanying global food crisis (2007/2008) by exploring how the rapid spread of soybean has transformed the landscape and the political economy of South American countries. Her main interest lies in how the politics of development in Latin America are conditioned by the changing structure of the global political economy.

[illegible] Agricultural Outlook 2021–2030 [illegible]

[illegible]

[illegible]

[illegible]

[illegible]

[illegible]

[illegible]

[illegible]

[illegible]

[illegible]

Part III
Emergent Issues

Chapter 8
Space, Scale, and the Global Oil Assemblage: Commodity Frontiers in Resource Peripheries

Michael John Watts

Abstract The purpose of this chapter is to make use of this expansive notion of extraction to explore the multi-scalar aspects—the socially produced space—of the oil and gas in resource peripheries. The scope of such a task is of course vast. As a consequence, my focus is a limited and rather unusual entry point into the global oil assemblage, namely oil theft (sometimes rereferred to as illicit bunkering). My goal is to use the illicit underbelly of planetary oil and gas as a way of exploring how scales are produced, how they intersect and overlap, and the tensions and politics associated with particular scales and with scale colliding or overlapping. In this story, the (petro) state figures centrally: not only in terms of the state itself as a sociospatial configuration engaged in the production of 'matrices of social space' that enable the extension of power and control and enabling the circulation of oil capital but also in terms of the rescaling of state territorial power that undergirds, and drives, globalization of the oil industry. The illicit aspects of the global value chain shed considerable light on how the operations of the oil assemblage entail a reconfiguration and re-territorialization of superimposed spatial scales, and not as a mono-directional implosion of global forces into sub-global realms; the relation between global, state-level and urban-regional processes can no longer be conceived as one that obtains among mutually exclusive levels of analysis or forces. The multi-scalar operations of the oil and gas assemblage—what is typically seen as the largely smoothly running global supply chains—exposes how the licit and illicit, the regulated and unregulated, and the ordered and disordered are operating together in tandem and often through the same mechanisms, infrastructure, and channels. Scale is multiple, overlapping, and often clashing replete with their own political orders and forms of authority. These are the circulation struggles—the imperatives to control place, space, and territory and what moves through and across it—which do not produce a clean logistical space, a well-ordered supply chain in which place has been thinned out or eviscerated. Quite the reverse. Forces of calculability and order fulfill "disorderly" functions and vice versa. They are organically and dialectically related and constituted. Logistical and infrastructural orders can be and regularly are disrupted, blocked, diverted, and appropriated in novel and creative ways, and all of this points to the co-production of

M. J. Watts (✉)
Department of Geography, University of California, Berkeley, USA
e-mail: mwatts@berkeley.edu

F. Irarrazaval and M. Arias-Loyola (eds.), *Resource Peripheries in the Global Economy*, Economic Geography, https://doi.org/10.1007/978-3-030-84606-0_8

logistical and political orders. Making things move and circulate is both an expression of power while constructing and depending upon systems of public and private authority.

Keywords Space · Scale · Oil assemblage · Hyper-extraction · Oil theft · Rent

When the drill bored down toward the stony fissures
and plunged its implacable intestine.
into the subterranean estates,
and dead years, eyes of the ages,
imprisoned plants' roots.
and scaly systems.
became strata of water,
fire shot up through the tubes.
transformed into cold liquid,
in the customs house of the heights,
issuing from its world of sinister depth,
it encountered a pale engineer.
and a title deed.
Pablo Neruda, 'Standard Oil Co.,' *Canto General*, 1940.

[T]he planetary mine…transcends the territoriality of extraction and wholly lends into the circulatory system of capitalism and now transverses the entire geography of the earth….[I]t provides important analytical insights for elucidating the role of rent, primitive accumulation and extra-economic force under contemporary capitalism.
Martin Arboleda, *Planetary Mine*, 2020.

8.1 Space, Scale, and the Oil Assemblage

The removal of oil from the earth, like mineral extraction in general, entails deeply spatial and territorial processes, and therefore understanding of the world of oil and its operations necessarily entails questions of scale. The wellhead, the concession, the oil host community, the pipeline network, and the oil state, each points to complex nested, overlapping multi-scalar character of the global oil assemblage. Neruda's famous poem takes us from the seabed and the world of oil's subterranean estates to the customs houses, the engineering companies, and the holders of title deeds. There is, of course, something irreducibly local and place-based about oil: the reservoirs are here and not there, the oil arises to the surface on floating rigs (say in the deep waters of the Gulf of Mexico) or onshore oil derricks (or in the Bolivian semi-arid *chaco*). But oil is at the same time global, wide-ranging, omnipresent, and capacious in its reach. It is oil's ubiquity and how it seeps into every aspect of our modes of living, moving, thinking, and dwelling—mobility, agriculture, pharmaceuticals, and plastics—that gives oil is unique and special qualities. Oil is, as Martin Heidegger (1963, p.93) suggested, as much an ontology as a material commodity: nature converted into a 'giant gasoline station'. It is a 'petroculture' as the *After Oil* collective (Petrocultures Research Group 2016) puts it, lubricating a post-industrial society in which fossil capitalism permeates and shapes values, practices, habits, belief, and affect (Szeman 2019). In this vein, a new book by historian Darren Dochuk entitled *Anointed with Oil* offers the view that the twin processes of Christianity and crude oil were foundational in the making of modern America. Oil is, in this accounting, a resource vested with extraordinary, even mythical, powers. It is built into the material world—the infrastructures of automobility or housing, for example—and the world of business behemoths and corporate power ('big oil') as much as it insinuates itself into virtually all aspects of politics, culture, and social life (Mitchell 2011; Huber 2012; LeMenager, 2014). Investing oil with these powers seems almost Althusserian in its forms of determination and structural force.

The idea that oil *is* power comes very close to being scorched by the fire of commodity determinism as Vitalis (2020) shows in his book *Oilcraft,* a form of modern-day witchcraft and magical realism. Oil is both extraordinarily visible—the power and reach of oil *capital* is not in question—and yet, says Vitalis, is marked by occlusion, exaggeration, the 'ideology of scarcity', and deception. Its powers are insidious seemingly everywhere but nowhere. It is an ecological catastrophe yet, as Szeman (2011, p. 324) puts it, "we do nothing about it" (see also Limert 2010), beyond our control while associated with a sort of depletion anxiety—'peak oil' is simply one of its monikers (Limert 2010; Kendall 2008).

Even some of the most compelling accounts of the oil landscape seem to suffer from a form of petro-myopia: Ferguson (2006, p. 204) famously described resource peripheries—he was specifically describing African oil zones—as 'enclaved mineral-rich patches' where "security is provided….by specialized corporations while the…nominal holders of sovereignty…certify the industry's legality….in exchange for a piece of the action". Ferguson offers a vision of oil seen through the lens of what one might call spatial mercantilism and the optics of "seeing like an oil company" (Ferguson 2005). The notion of an 'oil patch' and a mercantilist enclave characterized by social 'thinness' and a few local linkages overly localizes and insufficiently globalizes while diluting the complexity of the oil and gas assemblage. Seeing like an oil company ends up endorsing the Olympian powers of oil privileges in ways that need to be questioned. The first is the notion that oil capital gets what it wants from oil patches by, as it were, barely touching down through a sort of spatial confinement. And the second is the idea that the primary spaces of oil were locally and tightly circumscribed, as if all that mattered was the wellhead and corporate gated oil communities rather than the massive infrastructural, ecological and political economic imprints of the oil and gas global value chain. Seeing like an oil company privileges the 'sight' of the company for which its horizon and field of vision is the oil patch or enclave. But the enclave—or the hole in the ground—is just the starting point. If as Bridge (2009, p. 44) says, the hole is the necessary and 'essential feature of the extractive landscape', it must nevertheless be de-territorialized or, to use different language, rendered planetary (Labban 2014; Arboleda 2020). The oil well and the oilfield are planetary phenomena grounded in what Mezzadra and

Neilson (2019) call 'the operations of capital': the assemblage of oil extraction, logistics, finance, and corporate power.

Seeing like an oil company might be better rendered as a multi-scalar enterprise embedded in a global oil and gas assemblage. Big oil (national and international oil companies and so-called indigenous operators) is, of course, part of a global value chain and global production, distribution, and finance network (Tsing 2015). At its most capacious and expansive, this extractive assemblage also includes a suite of commodity trading houses, state actors, investment banks and different fractions of finance capital, engineering and service companies, shipping, refining, and other forms of logistics including state and private security forces and sophisticated systems of surveillance and control. And lest we forget, a heterogenous suite of other key actors and agents: oilfield insurgents, militias, local artisanal refiners, criminal organizations, trade unions, NGOs and advocacy organizations both local and global (Global Witness, Amnesty International), multilateral development institutions, development assistance agencies, and transnational regulatory institutions (one thinks of the Extractive Industries Transparency Initiative, EITI). To do justice to the oil assemblage is certainly to not only think in multi-scalar ways but it is also structured by, and exceeds, the operations of capital.

The idea of a vast and heterogeneous oil assemblage, replete with diverse actors and agents, exhibiting spatial complexity and the varied forms of territorialization, de-territorialization, and layered sovereignties that it entails points to a rethinking of contemporary extraction in relation to global capitalism in its various neoliberalized forms. Such reframing is what the concept of hyper-extraction (Shapiro and McNeish 2021; Aistrup et al. 2013; Biro 2002) is designed to address. In spatial terms—that is to say, with a full accounting of the production and management of oil and gas spaces—what hyper-extraction and the planetary well/mine point to is something akin to what Henri Lefebvre (2005; and Brenner and Elden 2009) calls 'territory': what he calls spatial hyper-complexity and nested, overlapping and fissioned spaces. The enclave space—perhaps less central than often thought—is one element of an oil and gas world constantly subject to forms of de- and re-territorialization, that is to say about a shape-shifting configuration of practices, flows, discourses, forms of authority, and politics operating across multiple scales.

Space and scale are here understood in quite particular ways, as socially produced or constructed. The spatial lexicon is redolent across the oil cosmos—the concession, the frontier, the enclave, the play, and the reservoir—and each of these spaces conjures up a stratified morphology, different scales understood as forms of spatial practice, representation (systems of signs and codes that are used to organize and direct spatial relations), and clandestine or subterranean spatial alternatives (spaces that the imagination seeks to change and appropriate). Scale is understood in a specific way (Howitt 1998), as both socially produced and as relational, rather than an abstract or as a naturalized category, as size (census tract, province, continent), or level (local, regional, national). Like environment, space, or place, scale is one of the elements from which geographical totalities are built (Swyngedouw 1997).[1] The challenge is to understand how particular scales become constituted, transformed, and reconstituted around relations of capitalist production, social reproduction, and consumption.

Lefebvre is especially useful because he sees the operations of the state—resource peripheral states for my purposes—at different scales as a response to, and a result of, political and economic restructuring that is a cause and consequence of globalization. Globalization entails capital flows of complex sorts, but it enrolls too the state whose interventions have identifiable and profound territorial implications. Lefebvre believes it is the role of the state to organize the spatio-temporal bases of the economic system at all levels of its operation. The globalization of capital, in its modern expressions, has required a significant reorganization of the world economy, which has been accomplished through the rescaling of state territorial power. Scaling and rescaling, territorialization and de-territorialization entail a double movement however. On the one hand, state and capital move together to produce and reproduce spatial hyper-complexity and, on the other, it—the multi-scalar assemblage—should also be seen as an opportunity for progressive political organizing against capital and the state (Marston 2000; Sayre 2017). Smith (1992) refers to such tensions as the 'politics of scale', where the territorial requirements of capitalism articulate extensions of power at the same time that these manifold scales provide openings to resist that power.

8.2 Petrolic Hyper-extraction in Resource Peripheries

Oil stands under or subtends the present only because it is also a uniquely sensitive region in the broader body of capital itself. Not only do oil companies occupy the commanding heights of contemporary economies, controlling immense swaths of Being, entire empires of material, land, and labour, but oil is itself preferred currency.....oil's financialization allows it to function as a speculative instrument bought to produce money directly out of money. It......simultaneously exerts an almost alchemical effect directly onto the body of all existing objects in the capitalist life-world.....

Andrew Pendakis, Oil and Being, 2017, p. 385.

Hyper-extraction can be construed in a number of related but distinctive ways. One is simply the expanded scale and output—the basic quanta—of resources extracted and consumed. From 1970 to 2017, the annual global extraction of materials[2] grew

[1] Swyngedouw (1997, p. 169) argues that scaled places are "the embodiment of social relations of empowerment and disempowerment and the arena through and in which they operate".

[2] Materials are taken to include biomass, fossil fuels, metal ores, and non-metallic minerals. Primary materials, sometimes labelled as raw or virgin materials, refers to materials sourced from mining and extraction activities in their raw form, such as mineral ores. These materials are entering into the economic system for the first time. Secondary materials refers to materials that have already been used previously (i.e. recycling). Materials extraction refers to the mass (physical weight) of primary materials extracted from the natural environment for use in the economy. See UNEP (2019, p. 42).

from 27 billion tons to 92 billion tons, while the annual average material demand grew from 7 tons to over 12 tons per capita, an annual average growth of 2.6%[3] (roughly twice the rate of population growth). The new millennium ushered in a major increase in global material requirements, which grew at 2.3% per year from 1970 to 2000, but accelerated to 3.2 percent per year from 2000 to 2017 largely driven by major investments in infrastructure and increased material living standards in East Asia and the Pacific. While there was a brief slowdown in the growth rate of demand for materials between 2008 and 2010 as a result of the global financial crisis, this has clearly had a very limited impact on the overall trajectory. Over the last century, resource extraction from non-renewable stocks has grown while extraction from renewable stocks has declined as the agricultural economy has contracted in relation to manufacturing.[4] Once accounting for some 75% of global material extraction, biomass today accounts for less than a third of total extraction. Global primary materials use is projected to almost double from 89 gigatonnes (Gt) in 2017 to 167 Gt in 2060 (OECD 2018).

There are other meanings of hyper-extraction too. One invokes speed, intensity, and energy (in the peculiar form of technological innovation) of contemporary extractive systems (Szeman 2017; Gomez-Barris 2017). The rate and scale of extraction is one attribute of its hyper-qualities—the massive scars and land movement entailed in the Canadian tar sands or Kennecott's Bingham Canyon copper mine—but there is relatedly the degree to which new technologies offer the possibility of enhanced recovery rates, the opening of new frontiers previously foreclosed (fracking is an obvious case), and the deployment of high-tech instruments for discovery, estimation, and surveillance of resources (three-D seismic, for example, in deep water mining or oil production). The very idea of the digital mine[5] or the digital transformation of the oil industry ("augmented reality, virtual reality, AI, intelligent automation, and the interconnectedness of all devices, hardware, and plant machinery will completely change the face of day-to-day oil and gas operations"[6]) are cases in point. The digital and the virtual point, naturally, to the particular sector-specific interfaces between extraction and infrastructures of various sorts, one expression of which is the sheer ability to move and transform and refine/process massive quantities of materials at unprecedented speeds and toward an array of novel end uses (rare earths and its role in the informatics sector is simply one instance) (see Klinger 2017).

Hyper-extraction also serves as a tagline for the gigantic, planetary scope and scale of the supply chain networks, not simply pertaining to the commodity extracted

[3] Resource extraction and processing make up about half of the total global greenhouse gas emissions and more than 90% of land- and water-related impacts (biodiversity loss and water stress).

[4] By 2010, non-renewable resource extraction represents over two-thirds of global material extraction with construction minerals making up over 30%, fossil energy 20%, and metal and metal ores 13%. Fossil fuels are the most traded primary material accounting for half of the global total of 11.6 billion tons of direct physical exports currently. Global primary materials use is projected to almost double from 89 gigatonnes (Gt) in 2017 to 167 Gt in 2060, OECD (2015).

[5] See https://www.miningreview.com/health-and-safety/the-digital-mine-how-miners-are-turning-a-vision-into-reality/.

[6] See https://www.oilandgasiq.com/oil-gas/news/what-is-digital-transformation.

but the density, connectivity, and tensions among different but functionally related supply chains—extractive, manufacturing, finance, and logistical—that intersect in extraordinarily complex global configurations (Bridge 2008). As a logistical order, and as a global supply chain, oil and gas is arguably one of the most vast, complex, and securitized of infrastructural and logistical spaces (Cowen 2010). In a manner unlike other natural resource-based global supply chains, oil and gas has (and has been since the early twentieth century) *the* exemplary case of a state-military-industrial-corporate complex. Like other infrastructures, oil and gas logistical systems are *unevenly* visible, they are both private and public (and sometimes hybrid mixes of both), and stand complexly in relation to spatial fixity (pipelines may be fixed but semi-submersible rigs are mobile between offshore fields in between periods of sedentary functioning). It is often said that large-scale technical systems are a system of substrates, invisible until they malfunction; they are taken for granted and to that degree offer up a sense (an illusion) of freedom. de Boeck (2012, np) says: "[Infrastructures] are mainly present in their absence". The Deepwater Horizon spill in the Gulf of Mexico is often taken as a case in point: offshore and out of sight to most American consumers.

Finally, there is the meaning of hyper-extraction as expanded, extended, or enhanced extraction. This more capacious view of extraction draws upon three strands of political economy. One is the move to de-territorialize and render planetary the mine most associated with the work of Labban (2014) and Arboleda (2020), and the idea that "capitalist urbanization secrets the planetary mine—everyday, above ground, scattered, diffuse, perpetual and swelling" (Labban 2014, p. 504). Central to the planetary approach is not simply scale, and interconnectivity (the city as the 'inverted mine') and breaking with methodological nationalism; rather, it is to see extraction as a set of shifting dynamic frontiers produced and enmeshed in forms of contemporary racialized capitalism and empire. The second is the related work of Sandra Mezzadra and Brett Neilson in their book *The Politics of Operations* (see also Mezzadra and Neilson 2017). Their focus is on the production of multiple edges and frontiers of expanding capitalism, the layered sovereignties and variegated legal spaces of global capital,[7] and the new spatial and temporal complexities of capitalism associated with capital's circulation and colonization of social life, what they call the politics of operations. In particular, it is the operations of a trifecta of 'sectors' and their connections that provide the core entry point: extraction, logistics, and finance.[8]

The third body of work has collectively addressed the question of contemporary capitalism and 'rule by rentiers' (Standing 2016; Mazzucato 2018; Piketty 2014) and for the purposes of this chapter, I shall refer to as extractive rents. Not surprisingly, financial rentiers, which is to say firms engaged primarily in financial activities and earning revenue primarily through the ownership and exploitation of financial assets,

[7] While acknowledging the importance of state sovereignty, they pull upon the work of Benton (2010) to emphasize the forms of quasi- or partial sovereignties, and the world of non-state world of petty sovereigns, to expose the fragmented and uneven complexities contemporary capitalism.

[8] In a very different register, albeit more sensitive to racialized extraction, Macaraena Gomez-Barris (2017) offers a decolonial theoretical account "foregrounding submerged perspectives" (p. 1) anchored in "anarcho-feminist indigenous critique".

have been in the spotlight, the principal agents of what has come to be seen as the dominance of Wall Street and finance capital. As a form of critique, rents is seen as 'unearned' (rather than productive as a source of accumulation), and owners of land, mineral resources, intellectual property, and a panoply of other income-generating financial and non-financial assets are seen to exercise a sort of hegemony within a neoliberalized and financialized capitalism. When economists refer to a rent-seeking political economy, they typically invoke a lack of market competition and hence the source of rent is state intervention or restrictions on economic activity. Others see rent as any income derived from ownership, possession, or control of assets (including financial assets) that are scarce or artificially rendered scarce. Implicit in differing explications of rent—too complex to enter here—is the notion of both monopoly power not only of ownership or control but also in the marketplace. In this sense, rent is income derived from the ownership, possession, or control of scarce assets under conditions of limited or no competition (Chrisptophers 2019, 2021).

Rent (and the rentier state) has been staples in the diet of extractive analysis, of course, for many decades (Yates 1996; Mommer 1990). Central to the rentier world so defined is the determination and distribution of property rights that are not deployed to produce new commodities but rather to extract value via rent (what has been called 'value' grabbing through 'pseudo-commodities' (Andreucci et al. 2017). There is, to take the idea of a planetary extractive system, an expanding class of rentiers operating in the interstices of the oil and gas assemblage (financiers, commodity traders, oil insurgents, politicians, military, corporations, and so on) who as it were profit without producing (Lapavitsas 2009). Rent-bearing assets—how they are created, their opportunities to extract value, and conflicts and struggles over the property rights that underlie them—are pivotal to contemporary capitalism, and to extraction in particular. Certainly, the state figures centrally in rents because (i) it typically creates and institutes property rights, (ii) regulates, enforces, and legitimates the distribution of rights and titles and their use, and (ii) because (and this is especially so in oil state), it is itself or acts like a landlord (a 'land appropriating state' or 'landlording state' (see Schmitt 2003; Hausmann 1981). But these rights might also inhere in international law or through multilateral institutions. Either way, as Andreucci et al. (2017, p. 38) put it, "the proliferation of private property relations over everything imaginable significantly expands the terrain for rent extraction and related struggles."[9]

Planetary extraction, and the dominant forms of neoliberalized finance capital associated with it, point to the centrality of the proliferation of rents and rent opportunities—'value grabbing'—to the operations of the oil and gas assemblage. This is no longer solely a product of corrupt rent-seeking petro-states but which operates across multiple spaces and sectors, across the licit and illicit, and among cores and frontiers, a development which has the effect of highlighting the blurring of conventional boundaries and borders in thinking about the global political economy of extraction.

---

[9] The proliferation of these rents means that they not only are the basis of capitalist expansion but are also the objects of contest and struggle.

The purpose of this chapter is to make use of this expansive notion of extraction to explore the multi-scalar aspects—the socially produced spaces—of the oil and gas in resource peripheries. The scope of such a task is of course vast. As a consequence, my focus is a limited and rather unusual entry point into the global oil assemblage, namely oil theft (sometimes rereferred to as illicit bunkering). My goal is to use the illicit underbelly of planetary oil and gas as a way of exploring how scales are produced, how they intersect and overlap, and the tensions and politics associated with particular scales and with scale colliding or overlapping (the politics of scale as Neil Smith put it). In this story, the (petro) state figures centrally: not only in terms of the state itself as a sociospatial configuration engaged in the production of 'matrices of social space' that enable the extension of power and control and enabling the circulation of oil capital but also in terms of the rescaling of state territorial power that undergirds, and drives, globalization of the oil industry. The illicit aspects of the global value chain shed considerable light on how the operations of the oil assemblage entail a "a reconfiguration and re-territorialization of superimposed spatial scales, and not as a mono-directional implosion of global forces into sub-global realms, the relation between global, state-level and urban-regional processes can no longer be conceived as one that obtains among mutually exclusive levels of analysis or forces" (Brenner 1997, p. 159). The multi-scalar operations of the oil and gas assemblage—what is typically seen as the largely smoothly running global supply chains—exposes how the licit and illicit, the regulated and unregulated, the ordered and disordered are operating together in tandem and often through the same mechanisms, infrastructure, and channels. Scale is multiple, overlapping, and often clashing replete with their own political orders and forms of authority. These are the circulation struggles—the imperatives to control place, space, and territory and what moves through and across it—which do not produce a clean logistical space, a well-ordered supply chain in which place has been thinned out or eviscerated. Quite the reverse. Forces of calculability and order fulfill 'disorderly' functions and vice versa. They are organically and dialectically related and constituted. Logistical and infrastructural orders can be and regularly are disrupted, blocked, diverted, and appropriated in novel and creative ways, and all of this points to the co-production of logistical and political orders. Making things move and circulate is both an expression of power while constructing and depending upon systems of public and private authority.

8.3 Moving Petroleum: Oil Theft Meets Oil Mafia

At peak prices, [illegally] tapping a Mexican pipeline of refined oil for only. seven minutes could earn a cartel $90,000.
Ian Ralby, Downstream Oil Theft, 2017

During April 18–19, 2018, the first global conference on oil and fuel theft, *Oil and Fuel Theft 2018,* was held in Geneva, Switzerland. Building upon the work of the

Atlantic Council (Ralby 2017; Soud 2020), the conference aimed to forge a global network of stakeholders in order to share information, expertise, and other mutual support in taking on what was referred to as 'a worldwide threat to security and prosperity'. *Oil and Fuel Theft 2018 drew* 140 attendees from around the world, including leadership of National Oil Companies (NOCs) from Iraq, Libya, Mexico, Ghana, and Uganda as well as government delegations from the United States and the Philippines and multilateral organizations like the World Custom Organization and the International Maritime Organization as UNODC and UNICRI, international oil and service companies and other corporate actors such as Dow Chemical. Among the offerings was some striking testimony by General Mahmoud Al-Bayati, Director General Counter-Terrorism and National Security Advisor for the Republic of Iraq, who outlined the history and genesis of how large-scale oil and fuel smuggling took root and in his country—coming to light in the infamous corrupt Oil-for-Food Programme between 1995 and 2003[10]—including the dynamics of oil smuggling for profit by ISIS in Iraq and Syria (Tichý 2019; Vienneast 2016). But Islam State and its oil investments are simply the tip of an iceberg, and these sorts of patterns could be repeated the world over. Oil theft is endemic in oil-producing states such as Nigeria, Mexico, Angola, Indonesia, and the Caucasus region.

The Atlantic Council's three reports—*Downstream Oil Theft: Global Modalities, Trends, and Remedies, Downstream Oil Theft: Implications and Next Steps*, and *Oil on the Water: Illicit Hydrocarbons Activity in the Maritime Domain*—offered the first comprehensive picture of global hydrocarbons crime. The scale of the illicit oil economy is mind-boggling. Globally, it is estimated that $133 billion worth of oil and fuel annually is stolen, adulterated, or fraudulently transferred at some point in its supply chain, an estimate that includes only refined (and not crude oil) products. But this figure itself is a massive underestimate since it does not include the sorts of losses associated with fraudulent oil trading contracts or oil revenues unaccounted for or 'lost' through public financial institutions in oil states like Venezuela or Nigeria. Liquified gas is also stolen and illicitly traded. Crucially, oil theft is not simply the preserve of petro-states in the Global South marked by 'poor governance'. In the European Union, the revenue loss caused by theft of oil and fuel is estimated to be worth 4 billion euros; the illicit cross-border trade in oil between Mexico and the USA involving not just Mexican cartels but American trading houses and oil companies is a multi-billion dollar business network (Jones and Sullivan 2019; Reinhart 2014). This illicit money machine turns on not only organized criminal gangs, terrorist groups, and insurgents but also corrupt public officials and security forces, offshore financial centers, and the global oil leviathan (the national and international oil majors). What

[10] The Oil-for-Food Programme (OIP), established by the UN in 1995 under UN Security Council Resolution 986, was established to allow Iraq to sell oil on the world market in exchange for food, medicine, and other humanitarian needs for ordinary Iraqi citizens without allowing Iraq to boost its military capabilities. Five reports were issued by the Independent Inquiry Committee concerning the United Nations Oil-for-Food Programme covering $64.2 billion in oil sold to 248 companies and in turn individuals and organizations sympathetic to the Iraqi regime, or those just easily bribed, were offered oil contracts through the Oil-for-Food Programme. See https://www.everycrsreport.com/reports/RL30472.html; and Jeong and Weiner (2012).

is on offer is a sort of global oil mafia operating in the interstices of the oil and gas global value chain.

Oil theft points, of course, to a larger systemic and structural pathology not peculiar to the vast oil and gas complex—according to market research by IBISWorld, the total revenues for the oil and gas drilling sector came to approximately $3.3 trillion in 2019, roughly 4% of global GDP[11]—but of the extractive sector in general: namely, endemic corruption and illicit financial flows (OECD 2016; see also Signe and Sow 2020).[12] In resource-rich post-colonial states, somewhere between 25 and 55% of global capital, flows may be illicit and it is widely acknowledged that illicit finance flows (IFFs) are deeply enmeshed with international crime networks (narcotics, arms, smuggling) and illicit commercial practice (tax and pricing fraud). Twenty percent of the 242 enforcement actions under the US Foreign Corruption Practices Act (FCPA) came from the extractives sector—by far, the highest for any industry, while of the 427 foreign bribery actions examined in a 2014 OECD report, 20% were lodged in the extractive sector as well (OECD 2014; Stanford Law School 2019).

The scale of the illicit financial flows in extractive economies across the Global South is gargantuan. According to Global Financial Integrity (2013), the real normalized cumulative IFFs from sub-Saharan Africa between 1980 and 2009 amounted to almost 1 trillion dollars (over $40 billion per year in the 2000s).[13] Net recorded outflows from West and Central Africa—and from the trio of oil producers, Nigeria, Congo, and Angola—swamped recorded transfers into other regions over the decade ending 2009. Oil and gas exporters accounted for over 55% of all IFFs in SSA over the same period. Data from the Brookings Institution estimates that between 1980 and 2018, sub-Saharan Africa received nearly $2 trillion in foreign direct investment (FDI) and official development assistance (ODA), but emitted over $1 trillion in illicit financial flows: four of the top seven African emitters of illicit flows 1980–2018 (totaling almost $200 billion) are oil producers (Signe et al. 2020). The latest trade data was reported to the United Nations to estimate the magnitude of import and export trade mis-invoicing—one of the largest components of measurable illicit financial flows—among 135 developing countries and 36 advanced economies (GFI 2020). By industrial sector, mineral fuels exhibited the second largest value gap ($113 billion, representing 16% of total trade) between the 135 developing countries and 36 advanced economies between 2008 and 2017. Over the same period, the average annual value gap in trade between SSA and the 36 advanced economies was $27 billion. For Nigeria alone, the annual estimated value gap between the country and 36 advanced countries was $4.8 billion.

---

[11] The assets of the largest ten oil and gas companies is roughly $3 trillion.

[12] "Corruption in commodity trading constitutes another emerging area of heightened risk given the substantial revenues diverted through this channel and their crippling effects on government budgets. Trade mispricing practices and complex kickback schemes to secure deals illustrate the increasing sophistication of constantly evolving patterns of corruption in this field", OECD 2016, p. 12.

[13] The report of the high-level panel on illicit flows from Africa, quoted by the United Nations Economic Commission for Africa (UNECA 2018).

Oil and gas provides the richest of soils for IFF risk (Gillies 2019). State-control of the industry in producer states is widespread and provides a massive hunting ground for rents on the part of the political, military, and businesses classes. The global supply chain is deeply financialized not only in the investment required for exploration and production but also most especially in the trading system, a domain marked by opacity and a lack of transparency on the part of the commodity trading houses, the transporters, and the investment banks. OECD's typology of corruption risks across the extractive sector investigated 131 major corruption cases (including oil and gas) noting that corruption arises at virtually any point in the extractive value chain (OECD 2016, pp. 11–12). The award of mineral, oil and gas rights, and the regulation and management of operations accounted for almost 75% of all cases, and involved bribery of foreign officials, embezzlement, misappropriation, and diversion of public funds, abuse of office, trading in influence, favoritism and extortion, bribery of domestic officials, and facilitation payments. Grand corruption, involving high-level public officials, is widely associated with the award of mineral and oil and gas rights, procurement of goods and services, commodity trading, revenue management through natural resource funds, and public spending. Sophisticated vehicles for channeling illegal payments, disguised through a series of offshore transactions, and complex layers of corporate structures often involving shell companies are recurrent features of the oil and gas sector landscape.

Perhaps no country on earth is more closely associated with large-scale oil theft (locally referred to as 'bunkering') than Nigeria—though Mexico, Iraq, and Russia follow close behind.[14] A report in 2018 declared the country to be the 'oil theft capital of the world'. The scale and costs of hydrocarbon crime in Nigeria are notoriously difficult to quantify not only because of the multiplicity of points where oil in its various expressions (crude, kerosene, refined petroleum, oil revenues) is stolen, but also because Nigerian state and regulatory authorities, as well as corporate actors, lack consistent and accurate metrics, and the commonly expected global standards for measuring and metering across the national supply chain are weak or entirely absent. Estimates of crude oil and fuel stolen and of revenues lost vary, often widely, as indeed does the data on pipeline sabotage and attacks. Nigeria lost approximately 204 million barrels, valued at N4.57 trillion (roughly $18 billion), to oil theft activities in the 4 years between 2015 and 2019, according to estimates by the Nigeria Natural Resource Charter (NNRC)[15]; that is to say, the Federal Government has lost approximately 43 percent of its revenue to oil theft over 4 years. Nigerian EITI estimated oil theft at $42 billion between 2009 and 2018,[16] and by some estimations 500,000 people are employed in the theft business broadly defined. According to international oil

[14] On August 29, 2019, the Ad-Hoc Committee of the National Economic Council (NEC) on Crude Theft disclosed that that Nigeria lost about 22 million barrels in the first six months of 2019. This loss was later put at $1.35 billion. This amount is already about 5% of the entire year's budget. Also, it is more than the capital allocations for education, health, defense, and agriculture combined. Yet this is crude oil lost in only one half of the year.

[15] See http://www.nigerianrc.org/nnrc-assesses-the-impact-of-crude-oil-thef.

[16] See https://www.nigerianbulletin.com/threads/nigeria-lost-n4-57tn-revenue-to-crude-oil-theft-in-4-years-%E2%80%93-maritime-security-review.413541/.

company figures, Chevron, Shell, and NAOC lost $11 billion between 2009 and 2011 due to theft and sabotage. The Nigerian National Petroleum Corporation (NNPC), the parastatal charged with the management of the industry, spent $2.3 billion on pipeline repairs and security from 2010 to 2012, and almost $1billion in the first quarter of 2019 alone.

Bunkering of Nigerian crude oil originated in the 1960s and late 1970s in part during the Biafran civil war but notably under military rule when the top army and navy officers began stealing oil—or allowing others to steal it—to enrich themselves and maintain political stability while also busting tight OPEC quotas. Local and foreign intermediaries did much of the legwork—Lebanese and Greek enablers loomed large—but the scale was small, perhaps a few thousand barrels per day. According to some reports (Katsouris and Sayne 2013, p. 5), lower global oil prices and Nigerian output, combined with the relatively closed group of actors involved, helped contain the business. Public claims that the Nigerian security forces—a Nigerian oil mafia—which were centrally involved in stealing oil grew after military rule ended in 1999 but by the 2000s the scale and scope of crude oil theft—in some cases reaching a staggering 200,000 bpd—grew to include elite members of the country's political and business classes: political 'Godfathers', businessmen, and well-placed political party members and high-ranking civil servants. As militancy arose across the oilfields in the 2000s, a new set of actors—insurgents, armed criminal and youth groups, local chiefs, and political operatives—muscled their way into the oil theft business. All of this pointed to a much more capacious system of oil theft—meaning the illicit capture of various assets and rents in the oil system, whether crude or refined oil, oil contracts and licenses, oil swaps, fraudulent trading deals, or simply stolen oil revenues that course through the country fiscal federal system. A government task force headed by Nuhu Ribadu (the *Report of the Petroleum Revenue Special Task Force*) reported in 2012 that up to 40% of oil products transported in pipelines between 2001 and 2010 were lost to theft or sabotage, and as much as 10% of the national production of crude oil were stolen amounting to N1 trillion annually ($5 billion). Not all oil that is 'missing' is necessarily bunkered—oil spills account for significant quantities each year[17]—Ribadu estimated that over a four-decade period, political elites had stolen $380 billion of oil and oil revenues.

8.4 The Illicit Life of a Barrel of Oil

Our story begins with a wellhead in the Niger delta, Nigeria. Nembe Creek Well 7, behind Mile 1 Community in Bayelsa State, feeds into the 97 km pipeline, Nembe Creek Trunk Line (NCTL). The trunk line is one of Nigeria's major oil transportation arteries that evacuates crude from the onshore fields to the Atlantic coast for export.

[17] According to government data (the national Oil Spill Monitor) from 2006 to 2019, a total of 12, 628 spill events occurred and a total release of 641,239 barrels. 75% of the incidents pertained to crude oil, accounting for 95% of total spill contaminants.

Owned by Aiteo Group, NCTL was recently purchased from Shell Petroleum Development Company (SPDC) as part of the related facilities of the prolific oil block OML 29. Construction of the NCTL commenced in 2006 and was finally commissioned in 2010 at the cost of $1.1 billion. Billed as a replacement to the aging and often vandalized Nembe Creek Pipeline which had suffered significant losses due to incessant fires, sabotage, and theft, SPDC made use of the pipeline to transport crude oil from the OML 29 starting at Nembe Creek to a manifold at the Cawthorne Channel field on OML 18, and finally to the Bonny Island oil terminal for export (and for liquified natural gas). With a capacity of 150,000 b/d at Nembe Creek, up to 600,000 b/d of liquids can be evacuated from the end point at Cawthorne Channel. In December 2011, barely one year after the line was commissioned, the pipeline was shut down for one month to repair leaks caused by crude thieves. In early 2012, Shell claimed that crude oil valued at $16 million (over 60,000 barrels per day) was being stolen *daily* from the NCTL.[18]

In global terms, the NCTL pipeline is arguably one of the most attacked, most sabotaged, and most compromised pipelines on the face of the earth.[19] It typically—that is to say, when it is not shut in by *force majeure*—seems to be losing more oil than it is actually transporting. Two short sections of pipelines in Brass and Nembe local government areas had over 600 attacks (due to theft, sabotage, and operational failures) between 2006 and 2019, and over 200 oil theft events in a 2-year period (2012–2014) (Whanda et al. 2017; Ngada and Bowers 2018). Within a year of opening, Shell discovered 17 bunkering (theft) spots along 3.8 kms of the pipeline.[20] Whether the interdictions are due to 'bunkering' (theft), sabotage, or operational failures, the collective assaults on the integrity of pipelines is simply staggering: according to government data (the accuracy of which is open to question both in terms of total numbers of spills and their cause), in the oil-producing Niger delta between 2006 and 2019 there were over 12,000 spill events, 75% of which were located in three states, Bayelsa, Rivers, and Delta and over 35,000 pipeline 'incidents'.[21] The official record says that over three-quarters of all spills and incidents were due to 'sabotage'.

Oil theft has waxed and waned since the return to civilian rule in 1999, shaped in part by the electoral cycle, by the price of oil, and by shifting patterns of criminal and militia activity. At some point, theft reached the staggering height of around 350,000 barrels per day and between 2006 and 2009, when an armed insurgency had thrown the oilfields into disarray, oil theft was funding the rebel cause (Adunbi 2015; Watts 2007; Naanen 2019; Obi and Rustaad 2011). Yet since the signing of the government

[18] https://www.ibtimes.com/nigerias-oil-thieves-drive-shell-distraction-company-plans-pipeline-sale-1558264.

[19] In 2018, four pipelines—the NCTL, the Trans Forcados pipeline, the Trans Niger pipeline, and the Obagi flow station—accounted for 600,000 barrels of lost crude oil.

[20] http://businessnews.com.ng/2012/02/07/nigeria-loses-over-500-million-monthly-due-to-crude-bunkering-shell/.

[21] The data is compiled by the Nigerian government National Oil Spills Detection and Response Agency (NOSDRA) and NOSDRA's oil spill monitor: https://oilspillmonitor.ng/; see also the Nigerian National Petroleum Company (NNPC) Annual Statistical Reports.

peace amnesty with 30,000 militants in 2009, oil theft has increased (Rexler 2019). In short, Well 7 feeds a rather leaky piece of oil logistical infrastructure. The oil black market, however, is just the beginning of a far more complex story.

8.5 The Social Life of the Illegal Tap

Let's follow that barrel of crude[22] as it crosses various spaces and scales from the wellhead into the trunk pipeline (the story may be slightly different if the pipeline is carrying refined fuels or gas). Within a short distance of the well, the crude flow is compromised by a 'hot' or 'cold tap', either on land or if the pipeline is running along the floor of the creeks and estuaries of the delta it could be underwater. Hot tapping involves creating a branch connection to a pipeline in which the oil is flowing under pressure. To access lines running underwater and to conceal the tap, a small area of swamp around the pipe may be cordoned off and drained. Either on land or underwater, an isolating valve is welded or fitted mechanically to the pipe. After fitting—and with the valve open—the pipe is drilled to the maximum permissible size through the valve, or the pipe is drilled part-way through and doused with sulphuric acid to complete the job once the line is in place. Exceptional skill and knowledge are required of oil infrastructure—the sparks from drilling can easily ignite the fuel—and tapping is typically undertaken by corrupt or former oil-industry workers (technicians and engineers) but increasing by a class of professional bunkerers who are part of 'unions' or small corporate groups. In the more elaborate (and large scale) bunkering, crude oil may be diverted from manifolds or flow stations[23] rather than individual pipelines, operations that require complicity and corrupt behavior from both local security forces and company operators.

In cold tapping, criminal or armed militias—sometimes referred to as oil gangs or oil mafia—blow up a pipeline, putting it out of use long enough for them to attach their spur pipeline. Many but not all pipeline bomb attacks appear to be linked to oil theft to enable a spur pipeline to be fitted but during the period of armed insurgency in the Niger delta (2005–2005), and earlier periods of activity by armed militias (2003–2004, for example), attacks and sabotage were either in retaliation for military operations against the oil gangs or as a way of extorting payments from transnational oil companies concerned to avoid *force majeure* (these were typically cash payments disguised as community development made to community 'youth organizations').

[22] In this section of oil theft, I make use of fieldwork conducted in the Nigeria delta over the last 15 years and the following: Katsouris and Sayne (2013), UNODC (2009), Oyefusi (2014), Fiennes (2020), Naanen et al. (2015), Schultze-Kraft (2017), and Ugor (2013).

[23] A manifold is a more complex arrangement of piping or valves designed to control, distribute, and typically monitor oil and are often configured for specific functions that require a higher degree of control and instrumentation. A flow station is usually the first stop for hydrocarbon fluids coming from crude oil and gas wells. Its purpose is to separate the hydrocarbon into liquid and vapor phases, reduce turbulence, and pass on the liquid to the next facility. The flow stations—there are almost 300 in Nigeria—may be located onshore, in swamps or offshore.

The illegal spur pipeline transports the crude, often over several kilometers, to a convenient creek, where it is released into flat bottomed loaders (barges) or wooden 'Cotonou boats' and then trans-shipped to differing locations, local, regional, and international. All stolen oil that is taken out of Nigeria for sale elsewhere—probably about 80% of all stolen oil—appears to be initially transported in surface tanks or barges, but much of the oil to be distributed within Nigeria. Depending upon the quantity, it is either sent to local refiners or to major state-run refineries, typically by trucks, and after refining by canoe to remote creek communities. The ability to tap with impunity requires the complicity of and the payment of rents by the tappers to local military and security forces (the Joint Task Force), local police, the coastguards, the security and low-level technical operators working for the oil companies. Local militants (they can be so-called secret societies, vigilante groups, ethnic militias, or anti-state insurgents such as the Movement for the Emancipation of the Niger Delta (MEND[24]) 'tax' the movement of stolen crude near to their creek encampments. Village chiefs and other youth groups in oil 'host communities' (there are roughly 1500 across the oilfields) impose imposts for permitting access to pipelines running across their community territory.

Arraigned in and around the tap and its installation, in short, is an ensemble of actors and agents held together by patterns of value-extraction and rent, at once forms of social life and political order: in short, a sort of ontology of infrastructures constituting an invisible supply chain (Ostensen and Stridsman 2017). The local tappers (skilled welders with experience in the industry as opposed to those hacksaw or puncture siphons) typically work in teams of 3–6 people and their proliferation particularly post 2009 and their networked relations with security forces and actors within the industry (the oil company community liaison officers, flow station technicians, oil service companies) have resulted in the rise of informal 'unions' (SDN 2017, 2018). The unions can arrange for a tap placement (a recent report says the fee is roughly \$6200—in a country where per capita income is \$2300) arranging often for the reduction in pressure on the pipelines by having company officials in the control rooms in the flow statins on their payroll (a reduction in pressure may cost a \$5–6000 fee). The unions provide security (paying off local security forces) and 'settle' with local community leaders. Operating the tapping points, irrespective of the scale of the tap, entails a 'consortium' of security (by local youth and payoff to local state and federal security forces), technical capability, and operational access capable of earning around \$1 million a month; the monthly costs entails a union

[24] The Niger Delta militant groups exhibit a suite of organizational forms (in which the lines between criminal, social, and political goals are often blurred), and among which the militants themselves may move. Some are so-called cults groups, such as the Icelanders and the Greenlanders, particularly in urban areas like Port Harcourt while other have a strong ethnic or cultural (clan) cast. There are a dizzying number of independently named militant groups, the most prominent in the recent past have been the Niger Delta People's Volunteer Force (NDPVF), led by Dokubo Asari, and the Niger Delta Vigilante (NDV), led by Ateke Tom, the Movement for the Emancipation of the Niger Delta (MEND) associated with Tompolo, Ebikabowei 'Boyloaf' Victor-Ben, and Farah Dagogo and more recently (post-2016), the Niger Delta Avengers. See Courson (2016), Golden (2012), Hazen and Horner (2007).

fee ($500), security payoffs ($1200), and labor and equipment ($4500). Payments to local community leaders in the oil host communities and to company technicians constitute additional expenditures. The union, with protection by local armed youth groups, delivers the oil to barges that in turn either move offshore or arrange for local delivery to artisanal refiners. All along, this local tapping and transporting supply chain are points of value extraction through rents which may take the form of extortion as much as formalized bribes and rents.

All of the oil companies including the national oil company in theory provide surveillance and security to manage pipelines for reasons of safety (since the costs of spills and explosions in and around communities, farms, and fishing grounds are especially high) and security. Yes, the massive proliferation of what the companies see as sabotage and theft suggests that either these systems are weakly enforced, or there is widespread collusion. Along some of the major trunk lines, there are serial taps—in some sections, there may be literally 100 or more taps per year! (Ngada and Bowers 2018). Both the companies and the federal government, moreover, have made use of local unemployed youth (as an employment strategy for a massive wageless class of alienated and frustrated youth across the region) to protect pipelines. But this in turn, building upon longstanding grievance between communities and the companies, has simply provided yet more avenues for value extraction and rent-seeking. Since 2009 as part of the amnesty plan, moreover, the federal government essentially placed 30,000 amnestied militants on their payroll including massive handouts to military 'commanders'. This demobilization strategy also entailed provided security contracts ('surveillance contracts') to the most powerful commanders—and to oil host community chiefs and local contractors—to protect infrastructure. But the entire amnesty program was simple in the business of attempting to purchase peace and a degree of consent in the context of a widespread insurgency. Wracked by corruption, the created tensions and conflict between former military groups and their leaders, while in practice simply superintended over the entire criminal oil theft enterprise. These surveillance functions were in effect 'ghost workers' rarely carrying out any work while providing expanded opportunities to extract rents (SDN 2019)[25] all the while converting commanders into local 'businessmen' and indirectly funding private armies.

The world of rents, 'taxes', and extortion are not confined to the world of tapping, shipping, and security, but also extend to the environmental cleanup and restoration. In view of the overwhelming number of spills (whether due to sabotage or operational failures) and pipeline interdictions each year, both the corporations and the federal and state regulatory agencies are liable for spill cleanup and restoration. A number of different federal and state agencies have jurisdiction over the oil and gas

[25] SDN (2019) describes an archetypical surveillance contract or as follows: "through the network of relationships the pipeline surveillance contractor maintains across communities, he is able to neutralize such opposition by distributing 'royalties' to 'settle' with chiefs, elders, young people, and women's groups before work begins. These demands typically amount to 10–15% of the value of the work to be done. He keeps track of the total amount of money he distributes and the international oil company reimburses him, depositing that amount into his bank account, using payments euphemistically known as 'local content'".

spillage response system but for regulatory purposes, it is the Joint Investigation Process (JIV)—which entails the submission of 4 forms to the national regulatory agency (NOSDRA) and a similar JIV-independent assessment by the company—that offers inordinate space for graft (Amnesty 2015, 2018; SDN 2016). The framework under which such assessments are conducted—contained in the Oil Spill Recovery, Clean-up, Remediation, and Damage Assessment Regulations, of 2011 as part of the NOSDRA Act—requires that a joint investigation team (JIT), comprising of the owner or operator of the facility from which oil has spilled, community and state government representatives, and NOSDRA, be constituted immediately after an oil spill notification is made, which ordinarily should be within 24 h after the occurrence of the spill. The JIT is required to visit the spill site, and investigate the cause and extent of the spillage. Under regulation 40 of the Clean-up regulations, it entails the conduct of a physical evaluation of the soil and surrounding environment, in order to determine the impact of the incident, the proper remediation procedure, and in monitoring the remediation progress. The horizon for value extraction when oil companies, the state, and impoverished poor communities are brought together in a 'stakeholder process' are legion. Amnesty International noted that "[t]he process is heavily dependent on the oil companies: they decide when the investigation will take place; they usually provide transport to the site of the spill; and they provide technical expertise, which the regulatory bodies lack." (Amnesty 2013, p.14). Participatory involvement is relatively limited and tokenistic: very few members of the community are able to participate in the process; typically, the oil companies generally deal with chiefs—or those they designate—and male youth leaders. Not infrequently community representatives have been asked to sign incomplete forms and communities are frequently denied a copy of the JIV form, even after signing it. Individuals are frequently paid to sign a JIV and company contractors in turn pay to get the clean-up contract. For example, a 2011 spill in Bayelsa State at Ikarama at a Shell facility illustrates intersecting forces of lack of transparency, of an inadequate response system capable of effectively responding to conditions in the Delta, and the corruption of the JIV process itself (Olawuyi and Tubodenyefa 2018, p. 8). Pressures and payoffs exerted by the operators including threats by security agencies resulted in the coercion of the community stakeholders to acquiesce and agree to the attribution of sabotage even though the communities believed it was an operational failure.

The systemic nature of the failure in broad terms is depicted in the SDN analysis of 5 years of NOSDRA data—6333 oil spills between January 2010 and August 2015 (SDN 2018). Over 75% of the reports had no estimation of spill area recorded, no description of impact recorded, no spill stop date recorded, and no post-impact assessment. All of this points to the ways in which the JIV process is not simply non-transparent and flawed as a data collection mechanism, but again a way in which consent is purchased through bribes, cash payments, and extra-economic coercion. Even in cases where corporate liability for the spill has been determined, the compensation systems are mired in corruption (NACGOND 2014). Large payoffs are made to chiefs, and compensation rates do not reflect actual losses and costs. The rehabilitation process drawing together companies, stakeholders, contractors, and companies plays true to form. Most of the rehabilitation work is subcontracted by companies to local firms but it is not clear that remediation meets international standards. Typically, contracts are awarded to local village or state elites at a price, ghost workers abound (faux youth groups who are paid to be quiet), and contract fulfillment is weak (cleanup work is often not even completed). When Nigerian novelist Chinua Achebe says that the Nigerian state is like one big crappy family, he might equally well have been describing the operations of the spills response and clean-up system.

When viewed through the prism of regulation and surveillance, the oil theft, spill, and clean-up system is one in which the foxes (regulators, companies, military, chiefs, militant commanders) are deployed to guard the henhouse. The whole assemblage is a shadow world of bribes, intimidation, extortion, fraud, and illicit finance. In a sense, the federal military forces fulfilled an identical function protected in its case by a thin veneer of arresting (without necessarily prosecuting) low-level barge operations and large numbers of small artisanal refineries, while leaving the black market operations intact. As amnesty payments to former militants dried up, or were absconded with by the commanders, and as employments opportunities through government programs declined as oil prices collapsed in 2014, many of the former militants had incentives to turn to artisanal refining and expanded tapping of pipelines.

8.6 The Piratical World

Tapping—hot or cold—is only among a number of means to steal crude oil. There are others: One is 'topping up' at the export terminal, another social space market, the point of connection between national and world markets. Oil company employees can be bribed into allowing unauthorized vessels to load. Authorized vessels can be topped—filled with oil beyond their stated capacity—and the excess load sold on. Oil revenues can also be embezzled, or money made through the sale of export licenses, credentials and bills of loading, and so on. This 'white collar' branch of oil theft allegedly involves pumping illegally obtained oil onto tankers already loading at export terminals, or siphoning crude from terminal storage tanks onto trucks. Bills of lading (B/L) and other shipping and corporate documents may be falsified to paper over the theft. Some topping off may also happen at sea—ship-to-ship transfer when larger barges holding up to 3000 metric tons of oil unload onto smaller tankers with a capacity of 10,000 metric tons anchored just offshore. Thieves generally use these small tankers to store and transport oil locally, though a few of the more seaworthy vessels may carry stolen oil to refineries or storage tanks within the Gulf of Guinea. Several small tankers can service a single oil theft network. Once the crude stored in them builds to a certain level, crews will transfer it to a coastal tanker or an international class 'mother ship' waiting further offshore. These ship-to-ship (STS) operations—typically occurring at night—can involve topping up a legal cargo of oil or filling up an entire mother ship. Oil theft from export terminals clearly entails not only a different set of actors from within the upper echelons of the industry but also a set of international agents—the shipping companies and a network of commodity traders and financiers—who can arrange for the international transfer and sale of oil products in China, North Korea, Israel, and South Africa. Political actors have a key role in the piracy networks, "due to their formal role in Nigeria's economy, as government regulators of the oil and maritime industries in Nigeria, or as businesspeople who process oil, provide support services to oil firms, and ship oil" (Hastings and Philipps 2015, p. 573). They are enablers and intermediaries standing between local economic and political networks and international actors operating in the global oil assemblage.

The other means is piracy constituted through other forms of rent extraction and different offshore, maritime spaces. The Gulf of Guinea, on west Africa's southern coast, and Nigeria's coastal waters, in particular, has become the world's most pirate-infested sea (Lopez 2015); Jacobsen and Nordby 2015). The International Maritime Bureau (IMB) reports that attacks on vessels at sea between Ivory Coast and Cameroon have grown dramatically since the early 2000s. Piracy has been common in Nigerian coastal waters over the last two decades but the region has a booming oil theft and kidnapping-ransom economy while in other piracy hotspots (Somalia, southeast Asia) piracy is in decline. Niger delta-based piracy has a historically deep pedigree dating back to the nineteenth century, but since the amnesty of 2009 pirates have the wind in their sails. Certainly, the number of attacks has ebbed and flowed this century, reaching an earlier peak in 2008 and 2013 (Nwalozie 2020),[26] but the current wave of violence is greater in scope and deadlier.[27] The number of crew kidnapped in the Gulf of Guinea increased more than 50% from 78 in 2018 to 121 in 2019.[28] Currently, the Niger delta region accounts for the vast majority of global maritime kidnappings: it equates to over 90% of global kidnappings reported at sea with 64 crew members kidnapped across six separate incidents in the last quarter of 2019 alone. The region accounted for 64 incidents including all four vessel hijackings that occurred in 2019, as well as 10 out of 11 vessels that reported coming under fire. As in South-East Asia, pirates in Nigeria used to confine themselves to raiding oil tankers to sell their cargo on the black market. When the oil price fell after 2014, they began copying their Somali counterparts and focused on kidnapping crews though oil theft made a comeback in 2018 and 2019. Unlike the Somalis, west African pirates rarely retain the vessels or the workers. Instead, armed with AK-47s and knives, they storm a ship, round up some of the crew, and return to land, where they hide their hostages.

Alternatively, if the prize is oil—and large quantities of oil pirates that cannot be trans-shipped to the coast—then pirates engage in ship-to-ship oil transfer to a mother ship. Again, different arrays of actors and rents are implicated. Pirates themselves

[26] According to one study Nwalozie (2020), actual and attempted pirate attacks declined in the wake of the 2009 amnesty but have grown since 2011 will a decline in 2014 and 2015 following the oil price collapse.

[27] The scale of the problem was such that in August 2019, the Nigerian Maritime Administration and Safety Agency (NIMASA) opened the Deep Blue Project's Command, Control, Computer Communication and Information Center at Kirikiri, Lagos.

[28] In 2019, over 90% of globally reported kidnappings and hostage-taking at sea took place in the Gulf of Guinea, and the vast majority of attacks are launched on shipping from within Nigerian territorial waters.

do not have personal access to the networks with which to profit from the oil and typically deliver the oil to the principals for a flat sum. Once loaded to the tanker, the pirate groups are directed by the broker to deliver to specific locations along the West African coast and to oversee security while loading to tanks on shore. In the case of the attack on the MT Anuket Emerald in 2012 by the MT Ejenavi, the oil was offloaded to the MT Grace, a chartered vessel, and then further transported to tank farms owned by local oil companies. At least one company provided brokerage services between the pirates and one of the oil companies that involved the MT Grace offloading the oil to the tanks of Integrated Oil and Gas Ltd., a company owned by the former Nigerian Minister of the Interior, Emmanuel Iheanacho. As Hastings and Philips (2015, p. 572) show, the boundary between licit and illicit has dissolved; the entities purported providing security are also involved in facilitating theft and providing protection: "the ship and cargo seizures are technically criminal activities, but at nearly every step of the way the pirates depend on the infrastructure (the ships, and storage and refining facilities) and the institutions (local brokerage, oil processing, and shipping companies, local and foreign buyers) of the formal oil economy". The visible and the invisible parts of the supply chain are in many respects indistinguishable: infrastructurally and functionally they are identical. To add another layer of complexity, the kidnapping and piratical networks often overlap and intersect with other illicit maritime networks in the Gulf of Guinea especially drugs, human trafficking, and commodity smuggling (Ralby and Soud 2018, UNODC 2005).

8.7 Afterlife of Oil: The Social Space of Illegal Refining

What is the afterlife to stolen crude? One answer is artisanal refining, locally known as 'Kpo-fire'. In virtually every community in the more isolated reaches of the Niger delta creeks and swamplands, households depended upon illegally refined fuels derived from stolen crude oil, typically selling at prices that undercut official fuel prices (Gelber 2015; Ikanone et al. 2014; Garuba 2010; Davis et al. 2007). Plastic jerry cans of artisanal fuel (kerosene and petrol) are ubiquitous, retailing at roundabouts and markets even in large cities such as Port Harcourt or Warri. A small percentage of Nigerian crude is refined locally in refineries that are notoriously inefficient and typically lose vast quantities of money: over 12 months between June 2019 and 2020, the four state-owned refineries were idled and had operational losses of \$367. The year previously, they operated at 13% of capacity. As a consequence, virtually all refined oil products are imported and then sold at subsidized prices (\$0.48 per litre), a sort of vast 'permit raj' that was exposed in a House of Representatives report in 2012 that entailed illicit activity of totally \$6.9 billion, one of the most monumental cases of fraud in Nigeria's history (Sayne et al. 2017). A 200-page inquiry revealed underhand practices fueled a six-fold increase in spending on oil handouts between 2009 and 2011. Fuel subsidies, part of a decades-old program meant to keep fuel prices low for millions of ordinary Nigerians, increased by 700% over 3 years. A report by a House of Representatives Committee identified the shadowy Nigerian

National Petroleum Company—ranked the world's least transparent state oil firm—as the centerpiece of massive theft. The firm was single-handedly responsible for almost half of the siphoned subsidy funds and was "found not to be accountable to anybody or authority". Seventy-two fuel importers, some with allegedly close links to senior government officials, were also singled out. In one case, payments totaling exactly $6.4 m flowed from the state treasury 128 times within 24 h to 'unknown entities'.

If the oil import business represents another massive tranche of the system of oil theft—in which traders and 'brief case' companies fight over the rents—fuel shortages nevertheless abound, especially in remote delta communities, and diesel and kerosene are in short supply and at a premium. In impoverished creek communities in which there is a sense in any case that the state (through nationalization) has stolen 'their oil', oil theft business was able to facilitate the emergence of what has become over the last decade a major growth industry. Every year, the security forces claim to have destroyed literally hundreds and in some cases thousands of illegal refining encampments dotted across the creeks in all of the oil-producing states.[29] A report estimated that by 2018, 43,000 barrels of crude were refined locally each day from roughly 500 camps (SDN 2013, p. 11); in two states (Rivers and Bayelsa), it was estimated that between 2013 and 2018 the number of refineries increased five-fold (to 2500) driven in part by national fuel scarcity and a growing demand for diesel and kerosene, and also by new forms of investment associated with "informal business associations improved information sharing and coordination of the supply chain" (SDN 2018, p. 4). As profitability has increased, new investors are bankrolling the camps and the distribution system, and a greater share of stolen oil now ends up on the domestic black market (roughly 70%). The value of illegal oil products in these two states alone was almost $1 billion.

Illegal refining arose during the Nigerian civil war (1967–1970) among Biafran rebels cut off from fuel supply, but as oil theft began to proliferate in the 1990s and especially the early 2000s so did the artisanal refining. During the period 2003–2004 as armed militia activity intensified, largely in response to state violence and the use by politicians and so-called political 'Godfather's in the 2003 elections of armed youth groups and so-called cults to intimidate opponents, competing non-state armed groups financed their activities increasingly through oil theft and refining.[30] One leader, Asari-Dokubo, claimed that he had a tapped pipeline running to his compound and that his refined products—'Asari fuel'—were cheaper, better,

[29] According to the Nigerian Navy, 2287 refineries were destroyed between 2015 and 2019 with a peak of 1218 in 2017: https://guardian.ng/news/nigeria/nigerian-navy-destroy-2287-illegal-refineries/.

[30] By the mid-2000s, there was a vast proliferation of militant groups, so say nothing of criminal organizations operating in the creeks, many of who were knee-deep in the oil theft-refining business: the Niger Delta Militant Force Squad (NDMFS), the Niger Delta Strike Force (NDSF), the Grand Alliance, Niger Delta Coastal Guerillas (NDCG), South-South Liberation Movement (SSLM), Movement for the Sovereign State of the Niger Delta (MSSND), the Meinbutus, the November 1895 Movement, ELIMOTU, the Arogbo Freedom Fighters, Iduwini Volunteer Force (IVF), the Niger Delta People's Salvation Front (NDPSF), the Coalition for Militant Action (COMA), the

and more widespread in creek communities than commercial refined products. The bunkering territories are protected and indeed fought over while the security forces—the Navy and the Joint Task Force—simultaneously destroy illegal refineries while taxing their operations to ensure that the well-connected and wealthy refineries are protected. The rate at which new refineries, especially since 2009, outpace the rate at which refineries are destroyed is substantial.[31]

Illegal refining depends, of course, on crude oil tapped from the tapping 'unions' who in terms delivered (and sold) the crude, often by Cotonou boats, to remote creeks refining 'camps'. The distributors typically exclude middlemen and the vessels (and their work crews) could be owned by the tap owner, by larger refiners or by local transporters and vessel owners (SDN 2013). Distributors unload the crude into open-air pits or into so-called plastic Geepee tanks. An average camp might have 10–20 people of all ages and genders; it requires capital investments (storage tanks, a 'cooking oven', a cooling system, and systems of hoses and drums). The refining process (dangerous and devastating for the environment) deploys a simplified version of fractional distillation in which crude oil is heated, condensed, and separated. A camp operator (who may or may not be the owner) has workers, security, managers, and 'boatmen' in his employ. Tappers might earn $30 per day, boatmen $50–150 per day. Setup costs for an average camp might be $5–6000 and might generate $7–8000 monthly income.

The refining process uses a simplified version of fractional distillation (locally called 'cooking'), in which crude oil is heated and condensed into separate petroleum products, aspects of which were adapted from traditional gin and palm wine distillation. The refining process begins when the 'black' is heated in an 'oven', burning crude oil to start the distillation, a process that releases dense black clouds into the camp, which, if not kept under control by spraying water onto the fire under the oven, can cause explosions. Distillation is kept cool through cold-water pumps and storage tanks, but the risks are substantial and the immediate impact on the environment catastrophic. The illegal refining process yields diesel, petrol, kerosene, bitumen, and waste products.[32] The yields of each product depend on the refining methods and the geological properties of the particular crude. Camps vary by size and their ability to operate in relation to the security forces. Because of the risk of detection by the JTF, most local refining occurs at night, when oil companies and security organizations have embargoes on staff field movements. Some larger and better protected refineries cook throughout the day that signals both the degree of normalization of illegal refining and that their operations are untouchable due to their close relationships with local representatives of the security services.

Greenlanders, Deebam, Bush Boys, KKK, Black Braziers, Icelanders, and a raft of other so-called cults.

[31] In 2019, the Navy destroyed 378 refineries in two oil producing states (Rivers and Bayelsa), apprehended 72 illegal speed boats, and captured 275 oil 'thieves'.

[32] After refining 30 drums of crude, a refiner might produce 25 drums of diesel worth N250,000; six drums of kerosene worth N30,000 in the black market.

The distributors typically represent a different network of actors and like tapping, it is one of the most profitable activities (in part because of the risks involved) in the oil theft assemblage. As the cost of buying stolen crude oil is a fraction of its true market price, the demand for cheap illegally refined products is considerable in both local and national markets. Most Nigerian crude oil grades are heavily diesel-rich, but the quality of refined products varies widely leading some refiners to purify diesel by mixing it with kerosene to improve the quality and launder the illegal product prior to distribution—much of which is sold in small quantities by women traders.[33] Illegally refined diesel has become so intermixed with legal diesel distribution networks that it is impossible to say how far illegal products are being distributed, but there is a brisk trade in locally refined produce to other coastal states including Lagos. Locally, blended diesel is sold through pre-negotiated sales or along the roads or near filling stations and typically undercuts the official subsidized price of commercial fuels by 15% or more. All movement and circulation operate under the cover of the police, the navy and military forces, and other security apparatuses (Transparency International 2019).[34]

In quantitative terms the major outlet historically for stolen crude is the international market. Barges of various sizes and conditions move the crude from the creeks where pipelines have been tapped—or in the case of theft at the export terminal, simply add to the existing cargo in the tanker. Making their way downstream, pulled by tugboats, the barges meet awaiting tankers that, due to the topography of the Niger delta, can anchor close to the places where the major rivers—the Benin, Escravos, Forcados, and Ramos rivers—empty into the Atlantic. Generally, the vessels involved are typically in poor repair (but may cost from $50,000–75,000, far beyond the means of most local oil tappers), and may have been officially decommissioned. One 2006 report put the cost of the barges at 6–10 billion Nigerian naira ($45–74,000), with a capacity of up to 5,000 barrels. The vessels seized by the Joint Task Force (JTF), the interagency body tasked with fighting oil theft, during the course of 2008 had a storage capacity of between 84 and 12,000 tons and comprised 44 barges, 50 wooden boats, 58 tankers, and 56 surface tanks. More recent interdiction work suggests bunkering activity is spread throughout the Delta region and there have been a number of high-ranking naval and military officers found guilty of involvement in the oil trade.

[33] The usual mix is estimated to be at a ratio of 1:2 (illegal: legal).

[34] A recent Transparency International report (2019, pp. 4–5) documented military personnel demanding payments from illegal refineries in exchange for allowing them to operate have continued to surface. These payments are reportedly regular and scheduled, and non-payment is punished. Interviewees in Bayelsa state, for example, reported that after an illegal refinery failed to meet a deadline to pay an 'operational fee' of 4 million Nigerian naira (USD$11,000), military officers arrived on the site and opened fire, allegedly killing one person and demanding an extra 200,000 Nigerian naira (USD$550) for the delay. The next day, 1.7 million Nigerian naira (USD$5,000) was delivered to military personnel with a promise to pay the balance of 2.3 million Nigerian naira (USD$6,000) later. Standard 'tax' payments for each drum of product was 1000 to 2000 Nigerian naira (USD$3 to USD$6) and retailers of illegal oil products spend an average of 60,000 Nigerian naira (USD$167) on transportation 'settlements' for different security personnel, including the military, and police at road check points if trucks were deployed to move the oil (they pay a bribe of 2000 Nigerian naira for each drum of product).

Most sources believe that the chain from theft up to transference to oil tanker or local distribution is handled by the same gang but generally different units of the same group, whereas the operation of the oil tankers and marketing of the stolen oil overseas appears to separate entities. While there are dedicated security forces devoted to surveillance and monitoring in order to apprehend bunkers, in practice a few arrests are made, and even fewer are prosecuted; in some cases, the tankers and their cargos mysteriously disappear.[35] In 2003, Brigadier-General Elias Zamani, then commanding a Delta peacekeeping force, was asked whether oil was being stolen by local people, the security forces, government officials, or an international element. His reply was: "All." (UNODC 2009, p. 22; see also De Montclos 2012).

Moving crude and refined products after they leave Nigerian waters is complex and murky, and much of the story of where Nigerian oil ends up and how much remains is unclear. The actors and networks are different from those operating at tap points, or local trader middlemen and the markets are varied: Benin, Ivory Coast, China, Singapore, Brazil, and Eastern Europe. Ghana is suspected of being an important portal. The actual transactions involved, the financing and the end point of the oil theft proceeds, all collectively point to the existence of a baroque transnational shadow world, a global network populated by a larger number of well-situated Nigerian and foreign actors. At the very least, it entails buying and renting barges and small tankers, chartering mother ships, fueling vessels, paying protection money, and having the requisite connections with the commodity trading world and at key port locations.

Tracing stolen oil is virtually impossible for several reasons. First, buyers of Nigerian oil load their cargoes onto tankers carrying crude from other oilfields, or even other countries—a process called 'co-loading'. For example, a trader might send a larger tanker to Nigeria to lift a 700,000 bbl cargo of Abo grade crude oil which then travels to the Forcados terminal, where it picks up an additional 300,000 bbl of Nigerian crude for delivery to Europe. Single tankers also commonly carry multiple 'parcels' of oil owned by different parties. The resulting full tanker-load of oil is called a 'split cargo' and each parcel onboard would come with its own bill of lading. Co-loaded and split cargoes while perfectly legitimate provide opportunities for bunkerers to disguise volumes of oil stolen at a terminal or in the field as a legal co-load. Mixed tanker-loads of stolen and legal oil are also rebranded as split cargoes by forging a separate bill of lading for the stolen portion. Second, complicated international delivery routes could also hide stolen parcels. After leaving Nigerian waters, a mother ship carrying stolen crude can offload all of its cargo at a single refinery, offload parts of its cargo at different refineries, offload all or part of its cargo into storage, transfer all or part of its cargo STS to another vessel (ship-to-ship transfer, STS), or transfer all or part of its cargo STS to multiple vessels. Virtually, all STS transfers of stolen oil probably take place further out at sea.

[35] In January 2005, two Nigerian admirals, Samuel Babatunde Kolawole and Francis Agbiti, were found guilty by a court martial of helping to steal an oil tanker and trying to sell stolen oil to an international crime syndicate. The vessel subsequently disappeared and although found guilty, they were merely sacked and demoted but were not sentenced to prison. See UNODC (2009, p. 25).

Finally, export oil thieves blend stolen Nigerian crude with oil from other countries and with fuel oil produced in or outside Nigeria. A range of customers buy the adulterated goods that result once they are mixed onboard tankers or at sites onshore. Some are probably sold as bunker fuel for ships. And finally, there is the murky world of storage. Most traders place large amounts of oil into storage facilities around the world. This enables them to blend crudes, or hold them until a particular market improves. Most oil storage is on land, but some floats at sea. Due diligence and reporting regulations vary by location. Selling crude oil into storage can allow sellers to disguise the oil's origins in future transactions. For example, an unscrupulous trader could receive a consignment of stolen oil into tanks it owns or rents, then blend or break it into smaller parcels. New bills of lading would be issued for each parcel when it was eventually sold, making less-diligent buyers less likely to ask for an original bill of lading created in Nigeria. Many sales out of storage also happen on an 'outturn' basis in which a refiner pays for a parcel of oil that a ship pumped into the buyer's tanks, not the volume stated on any bill of lading. The fact that the ship was carrying more oil than stated on the original bill of lading is then mostly irrelevant. This makes outturn sales a potential vehicle for hiding illegal top-ups. Not least, the locations, sizes, and beneficial owners of crude oil storage facilities—many of which are trader-owned—are not well known outside the trading community. In these scenarios, the ship represents a key node in the interlocking spaces of illicit oil: points of transfer, mixing, blending, and movement.

It bears repeating that the oil theft industry is dynamic and unstable, and varies in form and scale between states. Quantities are hard to determine with any accuracy—some IOCs have claimed that they lose 605% of their own from swamp operations, and aggregate totals have been as high at 400,000 barrels per day stolen—but it is clear that there was a sharp increase between 2002 and 2004, and a massive expansion between 2009 and 2016. In part, the volatility reflects not only prevailing changing oil prices but also the electoral cycle, the shifting politics and security situations across states, and the shifting incentives for export versus domestic markets. A recent report by SDN (2018, pp. 20–29) noted, in this regard, important realignments in the theft sector. One reconfiguration is the proportion of stolen oil now devoted to the domestic black market—increasing from 20 to 80% over the last decade. This is doubtlessly related to the post-2014 and 2019 oil price collapses and the implosion of the domestic commercial refining sector—and of course burgeoning domestic demand for fuels. Profitability and efficiency increased dramatically between 2012 and 2018, and in Bayelsa State earnings per camp increased 20-fold in part because the entire refining sector was dominated by one large group (the Rivers State black market was much more fragmented); the monopolistic structure reflected the ability of the group to pay off the security forces who drove competitors out of business. In general tap, camp and distributor 'owners' (and their investors) have gained, while camp salaries have lagged. Nevertheless, even camp workers' salaries far exceed the salaries of teachers, police, and even low-level military. Over the period, camp profitability roughly tripled from 17% in 2012 to 60% and 62% in Bayelsa and Rivers states, respectively, and total earnings grew from an estimated £24 million in 2012 to £578 million in 2016/2017. The annual estimated earnings of the various security agencies across both states rose from £1.5 million in 2012 to more than £30 million 5 years later.

In sum, oil theft co-produced a political order which sustained a largely ethnic insurgency; it is also part of a local 'participatory' economy—a very dangerous one—in which petty sovereigns ran illegal refineries for the local market. Diverted oil is also part of a transnational business—a local and transnational oil mafia—linking the high-ranking military, politicians, security, and domestic and foreign oil traders and shippers. The oil companies were an active part of this mix: local-level employees often conspired with refiners and oil thieves, while corporate executives saw this rough and tumble supply chain (its liminal qualities) as the price of doing business. This assemblage seen through the illicit is in many respects identical to the licit, operating across spaces and scales, one expression of planetary spatial-hyper-complexity.

8.8 Making Oil Circulate

Oil theft operations in Nigeria—as they are everywhere—entail a logistical and political order to tap, circulate, and distribute a variety of hydrocarbon products to local, regional, and international black markets. The dynamic and changing shape of this assemblage—including elite political actors, youth groups, local, international, and state-owned oil companies, shipping companies, insurgents, military, and much more—is both secret and elusive but yet in some respects operate in broad daylight. The fact that the movement of tankers, or topping up, or illegal refining can often operate openly and indeed through formal channels speaks to the fact that the 'invisible' (informal/illicit) supply chain operates through the same channels and with similar actors as the 'visible' (formal/licit) global oil and gas supply chain. The same actors can and are involved in both sets of activities. The boundaries blur, and the functions overlap and intersect. Furthermore, the invisible supply chain has its own formality. In the same way that the mafia constitutes a particular sort of order—a set of forms and conventions and relations to state powers—so too does the oil theft assemblage have its unions, taxes, dues, settlements, and returns. There are also enforcement (extra-economic) mechanisms; and like the formal gas supply chain, oil theft entrepreneurs and actors respond to the market, security, and political signals. The oil theft industry has its own lexicon: foremen, tappers, sponsors, investors, buyers, and traders. If the illicit oil supply chain is in many respects co-terminus with the licit—there is considerable porosity between the two—this observation also questions the view that the resource curse is simply a reflection of the fact that at every step in the process from extraction to final export, "oil firms are potentially subject to rents extracted by local political actors, both at the national and local levels, and must pay them off or establish informal understandings with them—often they must do both" (Hastings and Phillips 2015, p. 572). This is both true and incomplete since oil firms—of various sorts—are not simply complicit but are active agents in not just the extraction of value through rents but also the reproduction of the entire system.

The licit and illicit systems of petro-capitalism are deeply imbricated and mutually self-sustaining, feeding off of each other and exhibiting remarkable stability over time even in the face of conflicts and violence.

The shadow economy remains elusive and incompletely understood. Elusive not only because of secrecy and complicities at the highest levels of the state and government, but also because of the incomplete picture of the oil theft enterprise. The fullest report (Katsouris and Sayne 2013) claims that the oil theft supply chain or organization is more cellular than hierarchical. If Nigerian politicians and the press like to speak of bunkering barons and kingpins, or to describe oil-theft rings as mafias or syndicates, they argue that "most export operations are probably not run by one person, family or ethnic group, and management tends to be more cooperative than based on command-and-control" (ibid., p. 6). But these surmises are rather inconclusive, and there seem to be mafia-like organizations operating at multiple levels. It is clear that there is considerable heterogeneity across among the 'cells' networks and their membership; they vary in the size and location of their operations, needs and political entanglements, and morph and shape-shift in relation to the shifting winds of access and influence. Actors' influence and positions may wax and wane (military commanders come and go), but there is "a common set of roles to fill…high-level opportunists, facilitators, operations, security, local transport, foreign transport, sales and low-level opportunists" (ibid., p. 6). If all of this sounds like a fractal landscape that constantly shape-shifts, that is for now at least as robust a generalization as we can make of this oil assemblage and the operations of capital.

My account so far is restricted to oil products stolen from nodes within the logistical infrastructure and various rents extracted around these operations. But theft within the oil and gas sector operates in other hydrocarbon domains that arguably are of equal if not greater significance as regards illicit proceeds. One area pertains to illicit financial flows around the awarding of oil licenses and bonus payments through the leasing and tendering process. Licenses are assets that are traded among the political and business classes and represent one of the least transparent aspects of the industry, and the most corrupt (Sayne et al. 2017). Another is sales and so-called 'first trades', namely NOC-buyer contracts and terms of trade (which I turn to next). Other contentious domains include revenue collection and distribution (royalties, taxes, and public financial management), public procurement (Longchamps and Perrot 2017; EITI 2015), and the tendering of contracts issued in relation to oil and oil-related activities to oil service companies, and so on (Sayne et al. 2017). These arenas are replete with value extractions—rents—of the sort I described in oil theft, and often on a vast scale.[36] There is an entire industry—the transparency and accountability associated with the Extractive Industries Transparency Initiative (EITI) and advocacy organizations such as Global Witness and the Center for Research on Multinational Corporations (SOMO)—devoted to documenting the scale of the graft and theft associated with these illicit flows in these other domains of the planetary oil assemblage.

[36] In 2008 Albert J. Stanley, a former executive with a Halliburton subsidiary (KBR), pleaded guilty on Wednesday to charges that he conspired to pay $182 million in bribes to Nigerian officials in return for contracts to build a $6 billion liquefied natural gas complex.

Here, property rights often resemble theft: oil prospecting and oil mining leases acquired by members of the political class and sold, or oil infrastructure or massive bribes are paid to secure mega-engineering contracts. And not least there is outright theft—pillage really—at the highest levels of leadership. During the late military period in Nigeria (1995–199), the stolen assets sent out of the country by President Abacha to offshore financial centers were simply vast (estimated at $5 billion), and the process has continued (especially in the period after 2009). The infamous Bien Mal Acquis case affair (BMA) involved a series of corruption scandals which emerged in oil and mineral-rich central African states in 2007[37] and more recently Luanda-gate[38]—to say nothing of the presence of so-called 'oilygarchs' in the Paradise Papers and Wikileaks exposures—are the most dramatic cases of the pillaging of state oil resources. While the theft involved turns on corrupt political elites, the role of the national oil companies—the black holes of any national oil sector—and international oil companies and trading houses is central to any understanding of the unfathomable scale of financial hemorrhaging from the public purse.

Nigeria's oil theft universe returns us to Lefebvre's observation on global capitalism and space. First, oil theft is constituted through a myriad of overlapping, nested and intersecting spaces (the system is deeply territorialized): from bunkering territories to oil concessions, to pipeline networks, to trade corridors, oil host community territories, military jurisdictions, and so on. All are more or less regulated and orderly, each some form of quasi-sovereignty and populated by its own petty sovereigns. It is a space of hyper-complexity and layered sovereignty. Second, Lefebvre referred to a particular form of what he called state capitalism to understand the growth of post-war global capitalism. What is on offer in Nigeria and the licit/illicit value chain is less a version of *pur et dur* neoliberalism than a variant of Lefebvre's state capitalism. Parts of the supply chain—the planetary oil network—may resemble unregulated free for all capitalism, but other sites and nodes along the chain do not resemble the operations of free markets and stable and enforced private property rights.

---

[37] The BMA case (also known as the 'Ill-Gotten Gains Affair') is especially instructive. In March 2007, the French civil society organization, Comité Catholique contre la Faim et pour le Développement (CCFD), published the report, *Biens mal acquis profitent trop souvent: La fortune des dictateurs et les complaisances occidentales.* The report enumerated 23 instances of kleptocracy, and CCFD estimated, quite conservatively, that USD $100–180 billion of assets had been diverted, often to Western countries, by 23 national leaders and their families, targeting three African heads of oil states: Denis Sassou Nguesso from the Republic of Congo, Omar Bongo Ondimba from Gabon, and Teodoro Obiang from Guinea Equatorial.

[38] Isabel dos Santos, the daughter of former Angolan president José Eduardo dos Santos and Africa's richest woman, has a reported worth of over $2 billion. According to the Luanda Leaks, as well as reports from Maka Angola and other sources, Ms. dos Santos and her husband earned some of their money, thanks to public contracts approved by her father's government and suspicious deals struck with state-owned companies. It is notable that Ms. Dos Santos was appointed as head of Angola's NOC, Sonangol, by her father in June 2016 and remained in place until she was removed by the current President in November 2017 (see https://www.icij.org/investigations/luanda-leaks/).

8.9 The Cartel Model of Oil Theft

Nigeria has no monopoly on oil theft: Russia, Colombia, Iraq, and the Caucasus are known to have significant losses especially in downstream fuel theft. The fuel smuggling trade is vibrant across the Turkish border to Syria, and in the eastern Mediterranean, there is flourishing smuggling of oil focused on Libya, Malta, and Cyprus. In 2018, a major oil theft occurred in a Shell refinery in Singapore, the company's largest refinery in the world. In Indonesia, there were 63 cases of oil theft from pipelines from one concession, the Rokan Block managed by Chevron Pacific Indonesia. And in Europe, pipeline theft grew from barely a few cases in 2010 to 150 in 2015. Mexico, however, represents an intriguing case as a sort of counterpoint to Nigeria both in terms of scale and organization.

Mexico is a major oil producer and exporter of oil accounting for 15% of exports and 20% of state revenues, and like Nigeria has a large, complex national oil company (Petroleos Mexicanos [PEMEX]) controlling the upstream and downstream sectors.[39] But oil theft (*robo de combustibles*), illegal oil traders (*huanicoleros*), and pipeline taps (*tomas clandestinas*) have grown from a cottage industry run by local gangs ('grimy outlaws and bandits' as one commentator put it) during the 1980s and the 1990s[40]; it has now grown into a massive industry in the hands of cartels and specialized *huachicolero* syndicates who violently compete for control over the trade. Centered on two 'Red Triangles' located in Puebla and Guanajuato—with secondary centers in Veracruz and Tamalaulipas—by 2019, 22 states reported oil theft (Duhaukt 2017; Sullivan 2012). In 2006, there were 213 illegal pipeline taps; by 2016, it had grown to 6873 (accounting for over $11.3 billion for the period 2009–2016). By 2018, the number of taps had almost doubled to a staggering 12,582 (Jones and Sullivan 2019). Of the 1,533 pipeline taps reported in 2016, 1,071—or 70% of the total—were located along Highway 150D that parallels the trunk pipeline for refined products from Veracruz (and its refineries) to Mexico City. Not only was PEMEX itself in crisis, but oil theft and the violence it generated in a country marked by a pre-existing cascade of homicides (35,000 in 2019) reached crisis proportions. Oil theft in fact became one of, if not *the,* defining features of the first year of President Andres Manuel Lopez Obrado's *sexenio* following his landslide victorying July 2018 (Harp 2018).

The Mexico oil theft assemblage reflects a quite different architecture from Nigeria, rooted in the political history—and the political settlement—of post-revolutionary Mexico. Refined products (gasoline, diesel, kerosene) rather than crude represent the illicit commodities that are trafficked, and the focus is on a massive

[39] Mexico has 16 marine terminals capable of receiving imported fuel, plus 74 storage facilities and over 8,800 km of pipelines. The imports flow mainly through the Pajaritos, Tuxpan, and Veracruz terminals on the country's Gulf coast.

[40] The so-called King of Gasoline Francisco Guizar Pavon was one of the prime movers and old school *huachicoleros,* beginning his career in 1993 after he was laid off by PEMEX. Using his connection in the company, he developed a large fuel theft industry which was ultimately captured by the drug cartels. He was murdered in 2020.

underground system of distribution (and to a degree larger scale refining) designed to undercut official fuel prices.[41] Energy reforms put in place by the Nieto administration (2012–2018) permitted oil prices to rise and incentivized *huachicoleros* to undercut the formal market system. More crucially while there are extremely porous boundaries between the military and security forces, the oil thieves (unlike Nigeria, the central players in Mexico are transnational drug cartels) came to oil theft late in their institutional careers (the 2000s) on the backs of the deepening role of Mexico after the 1980s in the global narcotics (cocaine and heroin especially) wholesale trade. In part because of the anti-drug policies on both sides of the border and the changing markets structures for drugs, the cartels diversified and moved into oil theft, for which their national and transnational trade networks could be easily repurposed (Correa-Cabrera 2017).

The territorial natures of the cartels, their constant fragmentation, and division as a result of the Mexican government's kingpin strategy which produced intra- and inter-cartel violence, and the geography of the PEMEX pipeline networks, meant in practice a ferocious and violent struggle between cartels and other subsidiary or independent fuel traffickers to control the fuel business. And not least, the fuel cartels—currently the fuel trade is dominated by Cartel de Santa Rosa Lima, the Cartel Jalisco Neuva Generacion, and the Los Zetas cartel (a splinter group of the Gulf Cartel)—used their pre-existing military capabilities to extort and threaten PEMEX workers (to access pipelines, refineries, liquified natural gas storage tanks, and even offshore rigs), secure protection from the military and the judiciary, and develop a national (and cross-border to the US) tanker distribution system quite unlike the Nigeria domestic black market. The Mexican cartelized theft system is marked by extraordinary violence even by Nigerian standards: cities like Salamanca which house a large refinery and the PEMEX Minatitlan Mexico City pipeline have been marked by extraordinary bloodletting and conflicts between the cartels and gangs and by period pipeline explosion involving hundreds of casualties.[42]

Reading Mexico's oil theft history against Nigeria certainly throws up obvious parallels in terms of state capture and the complicities between the state oil companies, security forces, the political classes, and the oil thieves. But the actors, processes, and differing political histories and political settlements in each petro-state shape the specific forms in which the oil assemblage operates and reproduces itself. Oil theft grew out of, and was captured by, drug cartels that were at the time a product of both the changing global character of the drug trade, the nature of the drug markets, and the declining powers of the then ruling party, the Institutional Revolutionary Party (PRI), which dominated Mexican politics from the 1930s until the 2000 election of the first opposition party president, Vicente Fox, of the National Action Party (PAN). Nigeria's theft grew out of a political settlement in which an elite cartel presided over

[41] As in Nigeria with some of the militants who steal and refine oil and supply local communities, the oil cartels often have gained popular support by provisioning cheap fuel, making gifts of fuels to poor communities to celebrate holidays, and by developing a *huachicolero* subculture that involves the adoption of Catholic saints.

[42] A January 2019 explosion in Tlahuelilpan, Hidalgo, killed 135 people, and injured hundreds more.

a provisioning system and a multi-ethnic federal system that, for complex reasons, fed popular resentments on the oilfields that result in the proliferation of armed non-state groups and ultimately an armed insurgency and amnesty. These histories color the oil theft assemblages while retaining family resemblances. Interestingly, in a way that has no obvious parallel in Nigeria, the election of left populist President Obrador unleashed a major assault on the oil theft sector not only to stop the loss of revenues but also to stabilize a crippled PEMEX, reduce the extraordinary violence, and provide a better investment environment for investment by IOCs. By mid-2020, it was reported that oil theft had decreased by 90%. In August 2020, 'El Marro', the head of the Cartel de Santa Rosa Lima, was captured by Mexican security forces.

Whether in Mexico or Nigeria, oil theft reveals powerfully how the intersection of logistical/infrastructural and political orders exhibits ontological forms, what Julian Reid (2006) calls logistical life. In this sense, taps, refineries, barges, and export terminals enable "the public as social formation, [as] realms of interaction and collective consciousness" (Chalfin 2014, p. 106). These logistical and infrastructural orders, as uneven and irregular as they are over space, create different opportunities and different kinds of space "because they create the thickenings of publics, and offer the possibility of assembling people or slowing them down" (de Boeck 2010, np). Oil theft is built in and through oil infrastructures and represents what one might call an 'oil cosmos': not a circumscribed enclave of social thinness but an entire lived world. As a measure of this cosmos, one only need to note that the very presence of oil infrastructure (a wellhead, a pipeline) confers an existential status on communities: when present in a community territory, a village or town or city neighborhood becomes an 'oil host community' which confers particular rights, rents, and identities. Of course, if there is something of the entrepreneurial spirit at work in oil theft networks, and of resistance too (popular appropriation by those who see their oil resources as having been taken from them), it is a world of violence, conflict, subterfuge, and precarity. It is an ambient and combustible world, a vast provisioning machine in the business of shaping human experience and social identities. It is a sort of sensorium.

8.10 Frontier Spaces in Resource Peripheries

As a hyper-extractive, multi-scalar assemblage, the oil and gas supply chain operates less across a frictionless, smooth, monochromatic abstract space (in the sense used by Henri Lefebvre in his book *The Production of Space*) than through a networked mosaic of more or less regulated, more or less ordered, more or less licit, more or less calculable nodes, sites, spaces, and scales. The operations of the global oil and gas sector, like many supply chains, expose the radical unevenness of pumping and moving oil products. Put differently, "the trappings of logistical giants in one place actually hinge on logistical work in utterly deregulated zones .... Elsewhere" (Schouten et al. 2019, p. 779). These de-regulated zones are archetypical resource or commodity frontiers (Moore 2017; Tsing 2005). To the petro-geologist, the frontier has a set of technical meaning of courses. It is a geological province—a large area often of several thousand square kilometers with a common geological history—which becomes a petroleum province when a 'working petroleum system' has been discovered. The play has its own unique reservoir properties, particular temperatures, flow characteristics, viscosity, and so on. As geological formations located in space, these plays are often at the margins and that open and close with the shifting horizons of the exploration and production process of the oil industry.

These frontiers understood geologically—the fracking fields of North Dakota, offshore Gulf of Mexico in deepwater, the remote fringes of Russian Siberia, or the oilfields of Angola or Colombia—appear otherwise from the perch of the operations of fossil capitalism. That is to say, as a form of social space oil frontiers are rough and tumble, what Tsing (2005) described as sites of spectacular accumulation and of the 'not yet' and 'unmapped'. That is to say, an oil frontier is a prime example of the politics of operations where oil capital hits the ground. Frontiers in this sense are social spaces (themselves constituted by other overlapping, nested spaces) within the global supply chain "beyond the sphere of routine actions of centrally located violence producing enterprises….[populated] by classes specialized in expediency whose only commitment was to preserve the order that made possible the profitable utilization of such expediency" Baretta and Markoff 2006, p. 36). Frontiers as social spaces do not necessarily conform to Tsing's much quoted definition of the frontier: she says frontiers are unpredictable, free for all, not yet mapped, unstable (2005, p. 78). In my view, this is not quite right: frontiers can stably reproduce, and their dynamics frighteningly predictable and ordered. Frontiers exist at the limits of central power where authority—and indeed the rule of laws and its forms of enforcement and oversight—is neither secure nor non-existent but "is open to challenge and where polarities of order and chaos assume many guises" (Baretta and Markoff 2006, p. 51). Their key attribute is institutional patchiness what James Ron describes as weakly institutionalized spaces not tightly integrated into core states. Frontiers must be defined precisely in relation to the uneven presence and power of both state and capital but as Grandin (2019) says of the frontier, the state often precedes it; authority, power, and institutions of all sorts are present in complex and differentiated ways. Put differently, the characteristic of frontiers everywhere is the circumvention of infrastructural and administrative grids of the formalized economy.

The world of oil theft and invisible-visible supply chains shows how across the space of planetary oil are all sorts of frontiers, not all of which are the product of, or located within, unruly and corrupt resource peripheries of the Global South. Oil theft I have tried to suggest is a hyper-complex, multi-scalar phenomenon extending from the local to the global, blurring boundaries between illicit and licit, state and capital, market and society. The world of oil trading shows so clearly (Blas and Farchy 2021) *that* part of planetary oil is no less subject to opacity and lack of transparency than oil theft or the byzantine operations of national oil companies. Commodity trading houses and trading desks, offshore financial centers in the Cayman Islands, Luxembourg, Bermuda, Hong Kong, the Netherlands, Ireland, the Bahamas, Singapore, Belgium, the British Virgin Islands, and Switzerland too are frontier spaces part of the great cosmos of oil theft. Oil frontiers open and close over time and space; they appear and disappear. In so doing, they are part of the circulation struggle: the imperative to control place, space, and territory and what moves through and across it. Frontier phenomena that populate supply chains everywhere are marked by institutional patchiness, overlapping and nested forms of power and authority, and are quintessentially multi-scalar. Sometimes, the frontiers may throw up alternatives—counter-logistics or even emancipatory political orders—but as often as not they are precarious, violent, and illicit. The infrastructural and political orders that operate across and through the planetary oil assemblage may operate in close proximity to, or in conjunction with, the state and capital but equally they may exist largely outside of it.

References

Adunbi O (2015) Oil wealth and insurgency in Nigeria. Indiana University Press, Bloomington

Aistrup JA, Kulcsár LJ, Mauslein JA, Beach S, Steward DR (2013) Hyper-extractive counties in the US: a coupled-systems approach. Appl Geogr 37:88–100

Amnesty International Publication (2013) Bad information: oil spill investigations in the Niger Delta. International Secretariat, Amnesty International Publications, London

Amnesty International Publication (2015) Shell's growing liabilities in the Niger Delta. International Secretariat, Amnesty International Publications, London

Amnesty International Publication (2018) Negligence in the Niger Delta: decoding ENI and shell's poor record on oil spills. International Secretariat, Amnesty International Publications, London

Andreucci D et al (2017) "Value grabbing": a political ecology of rent. Capital Nat Social 28(3):28–47

Arboleda M (2020) Planetary mine: territories of extraction under late capitalism. Verso Trade

Baretta S, Markoff J (2006) Civilization and Barbarism. In Coronil F, Skurski J (eds) States of violence. University of Michigan Press

Benton L (2010) A search for sovereignty: law and geography in European Empires 1400–1900. Cambridge University Press, Cambridge

Biro A (2002) Wet dreams: ideology and the debates over Canadian water exports. Capital Nat Soc 13(4):29–50

Blas J, Farchy J (2021) The world for sale: money, power, and the traders who barter the earth's resources. Oxford University Press, London

Brenner N, Elden S (2009) Henri Lefebvre on state, space, territory. Int Political Sociol 3(4):353–377

Brenner N (1997) Global, fragmented, hierarchical: Henri Lefebvre's geographies of globalization. Public Cult 24:10/1:135–167

Bridge G (2008) Global production networks and the extractive sector: governing resource-based development. Journal of Economic Geography 8(3):389–419

Bridge G (2009) The *hole* world: scales and spaces of extraction. New geographies, vol 2. Harvard Graduate School of Design, Cambridge, Mass, pp 43–48

Chalfin B (2014) Public things, excremental politics, and the infrastructure of bare life in Ghana's city of Tema. Am Ethnol 41:92–109

Christophers B (2019) The problem of rent. Crit Hist Stud 6(2):308–309

Christophers B (2021) Rentier capitalism. Verso, London

Correa-Cabrera G (2017) Los Zetas Inc.: criminal corporations, energy, and civil war in Mexico. University of Texas Press, Austin

Courson E (2016) Spaces of insurgency. PhD dissertation, University of California, Berkeley

Cowen D (2010) A geography of logistics: market authority and the security of supply chains. Ann Assoc Am Geogr 100(3):600–620

Davis S, Von Kemedi D, Drennen M (2007) Illegal oil bunkering in the Niger Delta. Niger Delta peace and security strategy working papers series, Port Harcourt
De Boeck F (2012) Infrastructure: commentary from Filip De Boeck. Curated collections, cultural anthropology. http://culanth.org/curated_collections/11-infrastructure/discussions/7-infrastructure-commentary-from-filip-de-boeck
Duhaukt A (2017) Looting fuel pipelines in Mexico. Issue Brief. Baker Institute, Rice University, Houston. https://www.bakerinstitute.org/files/11960/
EITI (2015) The EITI, NOCs and the first trade. EITI Secretariat, Oslo
Ferguson J (2005) Seeing like an oil company. Am Anthropol 107:377–382
Ferguson J (2006) Global shadows: African in the neoliberal world order. Duke University Press, Durham and London
Fiennes A (2020) Oil bunkering in the Niger Delta. Contemp Challenges Glob Crime Justice Secur J 1:1–7
Garuba DS (2010) Trans-border economic crimes, illegal oil bunkering and economic reforms in Nigeria. Policy brief series, no. 15. Global consortium on security transformation. http://www.securitytransformation.org/gc_publications.php
Gelber E (2015) Making and unmaking of a social world: oil infrastructure and hydrocarbon dealers in Nigeria's Niger Delta. PhD dissertation, Columbia University
Gillies A (2019) Crude intentions: How oil corruption contaminates the world. Oxford University Press, London
Global Financial Integrity (2013) Illicit financial flows and the problem of net resource transfers from Africa: 1980–2009. GFI, New York
Global Financial Integrity (2020) Trade-related illicit financial flows in 135 developing countries: 2008–2017. https://gfintegrity.org/report/trade-related-illicit-financial-flows-in-135-developing-countries-2008-2017/gfi-trade-iff-report-2020-final/. Accessed 1 May 2020
Golden R (2012) Armed resistance: maculinities, egbe spiritsand violence in the Niger delta. Department of Anthropology, Tulane University, New Orleans
Gomez-Barris, M. (2017) The extractive zone. Duke University Press, Durham
Grandin G (2019) The end of the myth: from the frontier to the border wall in the mind of America. Metropolitan Books, New York
Harp S (2018) Blood and oil: Mexico's drug cartels and the gasoline industry. Rolling Stone. https://www.rollingstone.com/culture/culture-features/drug-war-mexico-gas-oil-cartel-717563/
Hastings JV, Phillips SG (2015) Maritime piracy business networks and institutions in Africa. Afr Aff 114(457):555–576
Hausmann R (1981) State landed property, oil rent, and accumulation in Venezuela: an analysis in terms of social relations. PhD dissertation, Cornell University
Hazen JM, Horner J (2007) Small arms, armed violence, and insecurity in Nigeria: the Niger Delta in perspective. Small Arms Survey, Graduate Institute of International Studies, Geneva
Heidegger M (1963) Discourse on thinking. Anderson JM, Hans E Freund (trans) Harper & Row, New York
Howitt R (1998) Scale as relation: musical metaphors of geographical scale. Area 30:49–58
Huber M (2012) Refined politics: petroleum products, neoliberalism, and the ecology of entrepreneurial life. J Am Stud 46(2):295–312
Ikanone G, Egbo M, Fyneface FD, Oduma I, Evans E (2014) Crude business. Port Harcourt, Social Action
Jacobsen KL, Nordby JR (2015) Maritime security in the Gulf of Guinea. Royal Danish Defense College, Copenhagen
Jeong Y, Weiner RJ (2012) Who bribes? Evidence from the United Nations' oil-for-food program. Strateg Manag J 33(12):1363–1383
Jones NP, Sullivan JP (2019) Huachicoleros: criminal cartels, fuel theft, and violence in Mexico. J Strateg Secur 12(4):1–24
Katsouris C, Sayne A (2013) Nigeria's criminal crude: international options to combat the export of stolen oil. Chatham House, London

Kendall S (2008) Bataille's peak: energy, religion, postsustainability. SubStance 37(2):146–149
Klinger J (2017) Rare earth frontiers. Cornell University Press, Ithaca
Labban M (2014) Deterritorializing extraction: Bioaccumulation and the planetary mine. Ann Assoc Am Geogr 104(3):560–576
Lapavitsas C (2009) Financialised capitalism: crisis and financial expropriation. Hist Mater 17(2):114–148
Lefebvre H (2005) The production of space. Blackwel.l, Oxford
LeMenager S (2014) Living oil: petroleum culture in the American century. Oxford University Press, London
Limbert M (2010) In the time of oil. Stanford University Press, Stanford
Longchamp O, Perrot N (2017) Trading in corruption: evidence and mitigation measures for corruption in the trading of oil and minerals. U4 Anti-Corruption Centre, Bergen
Lopez Lucia E (2015) Fragility, conflict and violence in the Gulf of Guinea (Rapid Literature Review). GSDRC, University of Birmingham, Birmingham, UK
Marston SA (2000) The social construction of scale. Prog Hum Geogr 24(2):219–242
Mazzucato M (2018) The value of everything: making and taking in the global economy. Allen Lane, London
Mezzadra S, Neilson B (2017) On the multiple frontiers of extraction: excavating contemporary capitalism. Cult Stud 31(2–3):185–204
Mezzadra S, Neilson B (2019) The politics of operations: excavating contemporary capitalism. Duke University Press, Durham
Mitchell T (2011) Carbon democracy. Verso, London
Mommer B (1990) Oil rent and rent capitalism: the example of Venezuela. Review (fernand Braudel Center) 14:417–437
Moore J (2017) The web of life. Verso, London
Naanen B (2019) When extractive governance fails: oil theft as resistance in Nigeria. Extr Ind Soc 6(3):702–710
Naanen B, Patrick T (2015) Private gain, public disaster: social context of illegal oil bunkering and artisanal refining in the Niger Delta. NIDEREF, Port Harcourt
National Coalition on Oil and Gas Flaring in the Niger Delta (NACGOND) (2014) JIV policy brief. National Coalition of Gas Flaring and Oil Spills in the Niger Delta, Port Harcourt
Ngada T, Bowers K (2018) Spatial and temporal analysis of crude oil theft in the Niger Delta. Secur J 31(2):501–523
Nwalozie CJ (2020) Exploring contemporary sea piracy in Nigeria, the Niger Delta and the Gulf of Guinea. J Transp Secur. https://doi.org/10.1007/s12198-020-00218-y
Obi C, Rustaad SA (eds) (2011) Oil and insurgency in the Niger Delta: managing the complex politics of petroviolence. Zed Press, London
OECD (2014) OECD foreign bribery report: an analysis of the crime of bribery of foreign public officials. OECD, Paris
OECD (2015) Material resources, productivity and the environment. OECD, Paris
OECD (2016) Corruption in the extractive value chain: typology of risks. Mitigation Measures and Incentives, Development Policy Tools, OECD, Paris
OECD (2018) Global material resources outlook to 2060. OECD, Paris
Olawuyi D, Tubodenyefa Z (2018) Review of the environmental guidelines and standards for the petroleum industry of Nigeria. OGEES Institute, Afe Babalola University, Ado Ekiti
Ostensen A, Stridsman M (2017) Shadow value chains: tracing the link between corruption, illicit activity and lootable natural resources from West Africa. Working paper, Chr. Michelsen Institute, Bergen
Oyefusi A (2014) Oil bunkering in Nigeria's post-amnesty era: An ethnopolitical settlement analysis. Ethnopolitics 13(5):522–545
Pendakis A (2017) Oil and being. In: Wilson S et al (eds) Petrocultures: oil, politics, culture. McGill-Queen's University Press, Montreal
Petrocultures Research Group (2016) After oil. Calgary, PRG

Piketty T (2014) Capital in the twenty-first century. Belknap Press of Harvard University Press, Cambridge, MA
Ralby I, Soud D (2018) Oil on the water: illicit hydrocarbon activity in the maritime domain. The Atlantic Council, Washington DC
Ralby (2017) Downstream oil theft: global modalities, trends, and remedies. The Atlantic Council, Washington DC
Reid J (2006) The biopolitics of the war on terror. Manchester University Press, Manchester
Reinhart LB (2014) The aftermath of Mexico's fuel-theft epidemic: examining the Texas Black Market and the conspiracy to trade in stolen condensate, 45. St. Mary's Law J 49:749–786
Rexler J (2019) Black Market Crude. Kleinman Center for Energy Policy, University of Pennsys-lvania, Philadelphia
Ron J (2003) Frontiers and Ghettoes. Princeton, Princeton University Press
Sayne A, Gillies S, Watkins A (2017) Twelve red flags: corruption risks in the award of extraction licenses and contracts. Natural Resource Governance Institute, New York
Sayre NF (2017) The politics of scale: a history of rangeland science. University of Chicago Press
Schmitt C (2003) The nomos of the earth. Telos Press, New York
Schouten P, Stepputat F, Bachmann J (2019) States of circulation: logistics off the beaten path. Environ Plan D Soc Space 37(5):779–793
Schultze-Kraft M (2017) Understanding organised violence and crime in political settlements: oil wars, petro-criminality and amnesty in The Niger Delta. J Int Dev 29(5):613–627
SDN (2013) Communities not criminals: illegal oil refining in the Niger Delta. Stakeholder Democracy NetworK, Port Harcourt
SDN (2016) Improving oil spill response in Nigeria. Stakeholder Democracy Network, London/Port Harcourt
SDN (2018) More money, more problems: the dynamics of the artisanal oil industry in the Niger delta over five years. Stakeholder Democracy Network, Port Harcourt
SDN (2019) Pipeline surveillance contracts in the Niger Delta. Policy brief. Stakeholder Democracy Network, Port Harcourt
Shapiro J, McNeish J-A (eds) (2021) Our extractive age: expressions of violence and resistance. Routledge, London
Signe L, Sow M, Madden P (2020) Illicit financial flows in Africa. The Brookings Institution, Policy Brief, Washington DC
Smith N (1992) Geography, difference and the politics of scale. In: Doherty J, Graham E, Mallek M (eds) Postmodernism and the social sciences. Macmillan, London, pp 57–79
Soud D (2020) Downstream oil theft. The Atlantic Council, Washington DC
Standing G (2016) The corruption of capitalism: why rentiers thrive and work does not pay. Biteback, London
Stanford Law School Foreign Corrupt Practices Act Clearing House. Accessed 27 May 2019. http://fcpa.stanford.edu/industry.html
Sullivan J (2012) From drug wars to criminal insurgency: Mexican cartels, criminal enclaves and criminal insurgency in Mexico and Central America. Fondation Maison de sciences de l'homme, Paris. https://halshs.archives-ouvertes.fr/halshs-00694083/document
Swyngedouw, E (1997) Excluding the other: the production of scale and scaled politics. In: Lee R, Wills J (eds) Geographies of economies. Arnold, London
Szeman I (2011) 'Editors' column: literature and energy future's. PMLA 126:2
Szeman I (2017) On the politics of extraction. Cult Stud 31(2–3):440–447
Szeman I (2019) On petrocultures. West Virginia University Press, Morgantown
Tichý L (2019) The Islamic State oil and gas strategy in North Africa. Energ Strat Rev 24:254–260
Transparency International (2019) Military involvement in oil theft in the Niger Delta. Transparency International, Berlin
Tsing AnnA (2005) Friction. Princeton University Press, Princeton
Tsing A (2015) The mushroom at the end of the world: on the possibility of life in capitalist ruins. Princeton University Press, Princeton, NJ

Ugor PU (2013) Survival strategies and citizenship claims: youth and the underground oil economy in post-amnesty Niger Delta. Africa 83(2):270–292

UNECA (2018) Base erosion and profit shifting in Africa: reforms to facilitate improved taxation of multinational enterprises. UNECA, Addis Ababa

UNEP 2019 *Global Resources Outlook: 2019:* Nairobi: United Nations Environment Program.

UNODC (2005) Transnational organized crime in the West Africa region. UN Office on Drugs and Crime, New York

UNODC (2009) Transnational trafficking and the rule of law in West Africa. UNODC, Vienna

Vienneast (2016) An investigation into oil smuggling and revenue generation by Islamic State. Vienneast Ltd, London. https://globalinitiative.net/how-organized-crime-and-terror-are-linked-to-oil-smuggling-along-turkeys-borders/

Vitalis R (2020) Oilcraft: the myths of scarcity and security that haunt US energy policy. Stanford University Press, Stanford

Watts M (2007) Petro-insurgency or criminal syndicate? Rev Afr Polit Econ 144:637–660

Whanda S, Adekola O, Adamu B, Yahaya S, Pandey PC (2017) Geo-spatial analysis of oil spill distribution and susceptibility in the Niger Delta region of Nigeria. J Geogr Inf Syst 8:438–456

Yates DA (1996) The rentier state in Africa: oil rent dependency and neocolonialism in the Republic of Gabon. Africa World Press, Trenton

Chapter 9
Scalar Implications of Circular Economy Initiatives in Resource Peripheries, the Case of the Salmon Industry in Chile

Beatriz Bustos, María Inés Ramírez, and Marco Rudolf

Abstract The circular economy has become the latest proposal to reduce the effects of prevailing modes of production in the environment. Its premise—that by breaking a linear understanding of production and creating new economic niches the environmental impacts of an economy can be reduced to zero—has attracted numerous industries under the loop for their negative environmental effects. The scalar implications of such policies remain to be fully understood, particularly in what may imply for changing relationships between resource peripheries and consumption centers. This chapter explores the case of the salmon industry to examine these implications critically. The salmon industry is a global production network connecting production sites in the global south and north with markets in Europe, Asia, and Latin America. The ecological impacts of salmon farming are significant, and global environmental networks have pushed for increasing certifications schemes, which include circular economy actions. Thus, it becomes an interesting case to put in dialogue arguments from resource peripheries literature on disarticulations and ecological contradictions, with circular economy arguments on sustainability.

Keywords Salmon production · Circular economy · Frontier regions · Environmental policy

9.1 Introduction

The Chilean salmon industry started over forty years ago with the introduction of salmonid species in lakes and rivers of the Los Lagos Region, in a process of trial and error to commercially adapt a northern species to southern waters. The success story

B. Bustos (✉) · M. I. Ramírez
Departamento de Geografía, Universidad de Chile, Santiago, Chile
e-mail: bibustos@uchilefau.cl

M. I. Ramírez
e-mail: maria.ramirez@ug.uchile.cl

M. Rudolf
Universität Heidelberg, Heidelberg, Germany

F. Irarrazaval and M. Arias-Loyola (eds.), *Resource Peripheries in the Global Economy*, Economic Geography, https://doi.org/10.1007/978-3-030-84606-0_9

has all the ingredients to define it as a resource frontier: the appropriation of natural resources (water and coves), capital investment in infrastructure and technology to make it happen, state-driven policies to favor the expansion of the activity, and labor availability. The process radically transformed the landscape and livelihood of the region in significant ways (Barton 1997; Román et al. 2015).

Socially, it triggered a process of proletarianization of peasant labor (Alfaro Monsalve 2015; Planells Gutiérrez 2019) added to the migratory flow of workers from elsewhere in the country attracted by the new industry and the consequent urban growth in the main cities of the region (Barton et al. 2013) followed by the emergence of new tensions associated with modernity (Bustos-Gallardo et al. 2019; Gallo 2001; Mansilla Torres 2006; Miller 2018). Economically, it can be explained through the idea of glocalization (Fløysand et al. 2010): given the centralized structure of the economic system, the essential flows of investment, and wealth did not immediately translate into better-off communities. Besides jobs, there were no other direct transfers of wealth between the industry and the local communities.

Environmentally, the industry's record is poor. Several studies (Buschmann 2001; Jacobsen and Hansen 2001; Leon et al. 2007; Leon Munoz et al. 2019) have shown that the industry affects the environment in at least three ways: through deposition of faecal matter which causes eutrophication of marine water, through salmon escapes which affect native species, and through deposition of antibiotics and other elements in the water transforming its quality. The increasing number of extreme events due to climate change is accelerating and amplifying the occurrence and expansion of these effects. Between the ISA virus crisis of 2008 and the massive algae blooming of 2016, there have been numerous events of escapes and massive mortalities (SERNAPESCA 2020) leading to social unrest and adverse reaction to the presence and continuity of the industry in the region (Bustos and Román 2019; Mascareño 2018; Neira Flores 2018). Furthermore, the process of production itself faces ecological contradictions that affect the continuity of the industry (Irarrazabal and Bustos 2018). Hence, the industry itself has had to assume a direct role in developing solutions addressing the harmful effects of salmon production in Chile.

The industry and the State have been criticized for a slow reaction to addressing environmental problems associated with industry's mode of production. Only after the ISA crisis, the State implemented stronger regulations to this effect (Bustos-Gallardo 2013). Meanwhile, the industry engaged in certification schemes such as ASC (Aquaculture Stewardship Council) and the creation of Corporate Social Responsibility offices to improve their interaction with local communities (Cid Aguayo and Barriga 2016). Yet, the effect of the measures taken has been short in changing social perceptions of the role of the industry in the current state of the region, but more importantly, how firms acted upon the territories. Instead of inducing relevant changes in the industry, certification schemes are considered a way to enter foreign markets, a checklist to comply to satisfy consumers. More recently, the industry has begun to engage in workgroups and state-led effort to discuss the potential of implementing a circular economy (CE) perspective in their production strategies.

A CE is an economic system that replaces the linear flow of resources by a circular flow of resources through reducing, reusing, and recycling materials used in production and consumption processes (Kirchherr et al. 2017). It contributes to sustainable development by establishing an alternative economic system that can prosper and simultaneously protects the environment. The concept becomes appealing for salmon producers because wastes are no longer understood as by-products of their core business that must be disposed of, sometimes incorporating high costs, but instead become inputs for new business. Thus, ideally, revenues along the commodity chain are increased, while at the same time environmental impact and costs are decreased: a win–win policy.

Not only in the Chilean salmon industry, but around the world, the CE is gaining momentum on the agendas of policymakers and industries and is seen as a toolbox that can be applied in different contexts to achieve sustainable development (Geissdoerfer et al. 2017; Schroeder et al. 2019). Meanwhile, the CE has been criticized for its variety of definitions, which lead to different perceptions of the concept, specifically concerning its connection to sustainability and the systemic perspective (Kirchherr et al. 2017). Moreover, its implementation has been shown to be highly context-dependent and not necessarily contributes to sustainability (Govindan and Hasanagic 2018; Korhonen et al. 2018; Winans et al. 2017). Instead, the concepts' misuse as a new form of a growth-driven and profit-orientated economic system may only postpone environmental degradation, or even accelerate it (Rammelt and Crisp 2014; Zink and Geyer 2017). Hence, general assumptions about the CE—e.g., waste reduction, environmental benefit, or sustainable growth—and the necessity arise to interrogate the concept's implementation in the specific industrial and cultural context under consideration.

On that note, the Chilean economy has long being recognized as a resource periphery (Barton et al. 2008) and coincide with Hayter (Hayter et al. 2003) that is precisely in this kind of economy whom provides a better understanding of the challenges that global capitalism has to sort the current environmental crisis. The interconnection of places of production and consumption through ecological crisis has been widely debated as well through GPN literature (Bridge 2008; Dorn and Huber 2020; Gargallo and Kalvelage 2020; Swanson 2015) but not much has been said yet on the rationale and actual changes that newly introduced solutions, such as CE, trigger in resource periphery contexts. Particularly, because the CE's motto is to "close" the circle in linear economies, but says nothing about the scale at which it happens, meanwhile, what precisely characterizes resource peripheries is the disconnection between spaces of production and consumption, which is at the core of the ecological challenges they face.

In this chapter, we aim to investigate how the Chilean salmon industry implements CE, to identify upcoming challenges as well as to evaluate if the CE contributes to sustainable development, or if it merely relates to more efficient waste management in a continuously unsustainable industry. The arguments for this chapter come from ongoing research led by the authors and have consisted in 19 qualitative semi-structured interviews of between 30 and 90 min with CE-related actors from the salmon industry, local recycling and waste processing companies, CE related scientists, and the Chilean government. All interviews were developed during the 2019–2020 period.

In what follows, we will develop the following argument: the logic of the CE has fit well with a corporate culture of each on their own that has prevailed in the salmon industry history, and in other words, it has become a new paradigm to justify continuous expansion and growth. But, while it improves efficiency, it does not yet resolve the underlying problem of resource peripheries. Thus, in the current stage of the salmon industry in Chile, waste has become the new frontier of accumulation. This stage has led to the emergence of new actors (the firms and innovators leading the valorisation), who are demanding new responsibilities and a change in the rules of the game which may lead to a new culture and interaction with communities but is still too early to say. The chapter is structured in three sections. After this introduction, we provide a brief overview of the salmon industry in Chile delving into the ecological contradictions that put waste treatment at the centre of the challenges for the sector. Then, we develop the evidence on the implementation of CE in the Chilean salmon sector, and we end up with a section of conclusions, where we, based on the salmon case, provide a discussion on the challenges of CE for resource frontiers.

9.2 The Salmon Industry in Chile

The Chilean salmon industry today comprises over 40 producer firms from national and transnational capital (Japan, Norway, and Germany). Still, only 10 of them concentrate 90% of production (MOWI 2020) and nearly 66% of exports and the weight of the market (TERRAM 2018). Export destinations are mainly the United States, Japan, and Brazil, reaching nearly five billion US$ in sales in 2018, according to the Chilean Central Bank (Garcés 2018). Production of salmon in Chile can be understood as a resource frontier: growing exponentially, from 70 tons in 1981 to 630,000 tons in 2008 (SERNAPESCA 1998–2008) highly concentrated in the Los Lagos region, before the ISA virus shacked the industry and production plumbed to below 200,000 tons in the following year. Since then, the industry has expanded moving to the southern regions of Chile—Los Lagos (50% of production in 2018), Aysén (40%), and Magallanes (10%). Although farms have increased salmon production in Aysén over the last decade, the Los Lagos region is still the one concentrating on processing plants and providers. In terms of workers, the industry reports providing nearly 21,000 direct jobs and 35% of unionization[1] (SalmonChile 2020). According to a survey conducted by the National Statistics Bureau (INE 2018), Los Lagos concentrates 56% of workers in the pisciculture stage, 43.2% of the growing stage, and 75.8% of processing plants stage. Thus, salmon is an activity always

[1] The national rate is 20% (https://www.ciedess.cl/601/w3-article-3753.html#:~:text=Si%20se%20compara%20con%20los,promedio%20que%20es%20de%2030%25).

expanding and looking for the next place/technology that could provide an advantage in the accumulation process, its constant flux of people, capital continually reconfigures the landscape of accumulation in the salmon frontier.

Often portrayed as a success story, those first 20 years also contained moments of ecological crisis such as the algae blooming of 1988 (the brown tide) that caused significant disruptions in production and was followed by consecutive events in 2000–2002–2004–2005–2009–2011, being 2004 an event that caused severe damages in production (Luxoro 2018). Along with those events, the industry experienced an increasing presence of pathogens that lead them to excessive use of antibiotics (Cabello 2004; Fortt et al. 2007; Millanao et al. 2011) which in turn became an issue for consumer and environmental NGOs for its potential effects on native fisheries. Furthermore, the constant escapes that showed the fragile ecological equilibrium sustaining production. By 2016, the year of the last massive algae blooming, production had risen to 837,000 tons (Subpesca 2017), showing that even after the repeatedly occurring crises, the focus was on increasing production not on dealing with the environmental passives it generated. This section will expose how and to what extent waste became an issue for the industry, so the next section can discuss why the CE concept resounded with salmon firms.

9.3 Waste and Ecological Contradictions

Salmon farming is an economic activity highly dependent on the environmental conditions, which at the same time, is degraded with the advance of intensive development of aquaculture. Therefore, as an economic activity, it represents a series of ecological contradictions, partially addressed with sanitary regulations and new operational standards (Bustos-Gallardo and Irarrazaval 2016). The crisis of the ISA virus in 2007 (Bustos-Gallardo 2012), as well as the Harmful Algal Bloom (HAB) in 2016, have put the salmon industry to the test. Moreover, these incidents warned about a new barrier to production: waste management. Considering that the last HAB crisis exposed the lack of facilities suited for their treatment or disposal: from 40,000 tons of fish mortalities, 4,600 were thrown in the sea (Buschmann et al. 2016). As a result, the Province of Chiloé experienced an environmental, health, economic, and social crisis.

Salmon has different productive phases: pisciculture, smoltification, fattening, and processing. Each of these phases generates different types of waste. The pisciculture phase generates 16 types of waste, followed by processing plants (11) and finally the weight gaining and smoltification processes, with nine and eight, respectively. In all cases, sewage, liquid industrial waste, fish mortality, household, and hazardous waste are generated, as well as plastics, polystyrene foam, and cardboard (Stack Lara et al. 2018).

However, quantifying salmon farming residues in Chile is tricky, as the Chilean regulations for the reporting of residues are recent (2015) and the data aggregation does not allow the separation of salmon farming from other aquaculture and

Table 9.1 Quantification of the most representative waste in salmon farming 2011–2015. *Source* Stack Lara et al. (2018)

Waste	Amount	Unit of measure
Sewage sludge + industrial liquid waste	1,314	Ton/year
	109,512.23	M3/year
Domestic waste	1,117	Ton/year
	2,336.50	M3/year
Mortality	24,419	Ton/year
	1,220.93	M3/year
	128,152	Un/year
Plastics	11,144.461	Un/year
	28,500	Ton/year
	1,666.20	M3/year
Paper, cardboard and scrap	153.97	Ton/year
	545	M3/year
Hazardous waste	121.18	Ton/year
	3689	Un/year
	258.01	M3/year

fishing activities (Superintendencia_del_medioambiente 2020). Moreover, companies declare their waste through different metrics. A study calculated the waste, as indicated in Table. 9.1.[2] Adding only values expressed in tons and discounting the wastes that have some treatment standards (fish mortality, industrial liquid waste, and hazardous waste), close to 60,000 tons of waste go to landfills during the salmon production process every year. The ongoing accumulation of these large quantities of waste becomes a problem for the industry, as the "salmon regions" (Los Lagos, Aysén, and Magallanes) have a limited capacity to dispose of domestic and industrial waste, which will be illustrated in the following section (Fg. 9.1).

As can be seen in Fig. 9.1, one of the highest volume wastes is salmon mortalities. However, it is also one of the wastes that have specific regulations by the National Fisheries Service (last modification in Exempt Resolution No. 540 of 2020), unlike the inorganic wastes studied here. According to an official of the Undersecretariat of Fisheries and Aquaculture, the health contingency forced companies to increase their processing capacity in reduction plants inside farming operations, as well as

[2] The appendix of the report, containing the whole dataset used for the summary of waste numbers displayed in Fig. 9.1 was not public, and we obtained it through transparency law. We raise a word of caution because we found some inconsistencies in the reported figures but decided to provide them after revision. The report has so far been the only serious effort to quantify the waste problem. It is difficult to find data about salmon waste in Chile. The authors decided to use this report, instead of the provided from the Ministry of Environment, because the latter doesn't count with data segregation according to the needs of this article. Indeed, this register groups all aquaculture activities in one category: salmon, mussels, etc. For their part, the cited report from Stack Lara et al. (2018) was tasked by the Subsecretary of Fishing and Aquaculture in Chile, who was also contra part. Therefore, it is validated by one of the most important public institutions in the matter.

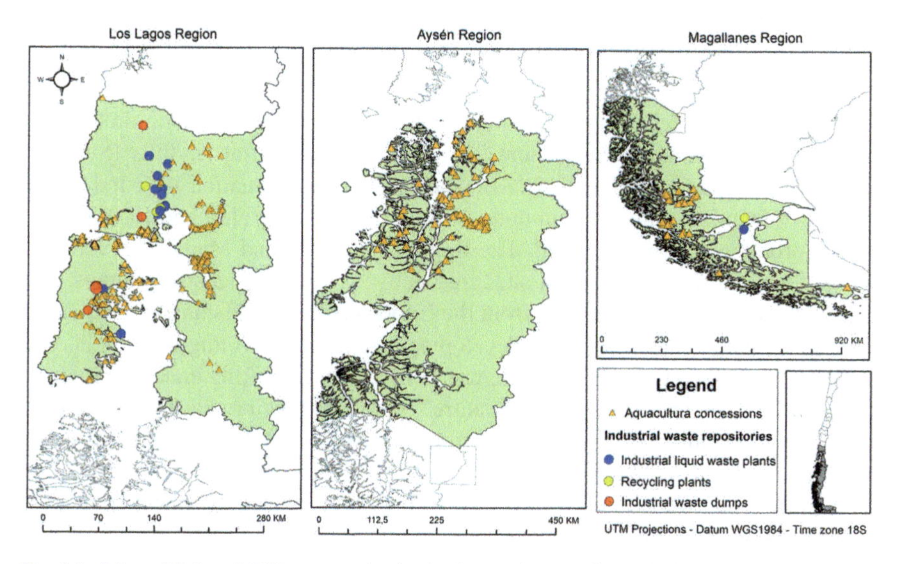

Fig. 9.1 Map of Industrial Waste repositories in the southern regions of Chile. Own elaboration

the magnitude of storage, in case of more mass mortality events in the future, where the culture centers must have: "Feasibility of extraction of 10 tons, denaturation feasibility of 15 tons and storage capacity of 20 tons" (Subpesca official 2020). The denaturalization process implies giving reuse to the residues, through procedures such as ensilage. This is used in the production of by-products such as oils and agricultural inputs (Induambiente 2012). The silage technique was used during the 2016 crisis in 57.12% of the total waste, which ended up in reduction plants, while 30.34% went to landfills (Seremi de Salud 2018). However, a remaining 4,600 tons of mortalities that could neither be ensilaged nor landfilled was disposed of in open seawaters.

Although storage capacities improve, evidence shows that mass mortality events exceed the capacities indicated by Subpesca. In fact, in the HAB mortalities that occurred in April 2021, only one farming center accounted for 1,300 tons for the event. On the other hand, an industrial landfill manager indicates that in the event of another mortality event "*the same thing will happen. There is no one to absorb that amount. And besides, it is reasonable. […] An industry cannot be prepared for an event that is going to happen once in one's professional life, I do not think it will happen again. At that level*". In this context, mass mortalities are both beyond the scope of the CE and beyond the capacity of waste management.

9.4 Waste as a Barrier to Accumulation

In the Los Lagos Region, only five domestic waste dumps are in operation (Undersecretary_of_Regional_Development 2018) and two industrial landfills (Superintendencia del Medio Ambiente 2020). The most critical situation is found in the Province of Chiloé, with the commune of Ancud on sanitary alert and all industrial dumps prohibited from operation. The Regions of Aysén and Magallanes have 19 and 8 domestic waste collection sites, respectively, and no industrial dumps (see Fig. 9.2). According to an official from the Undersecretary of Fishing and Aquaculture, this situation stems from the development of salmon farming in its industrial phase (1980s) when the issue of waste was not recurrent, working under the logic of what Moore (2015) has called "cheap nature".[3] Waste regulations have been reactive, with new or modified regulations appearing in response to the ISA or HAB crisis rather than in a preventive way.

Historical lack of action on this issue and poor management at the national level of industrial waste issues has led to the current scenario. Only in 2016, the parliament approved the Extended Producer's Responsibility Law 20920 (Law EPR) which holds producers responsible for the waste they generate. However, it applies to certain materials only, and until to date it has not been fully implemented. Furthermore, there are no specific regulations for salmon farming waste (Terram 2019) and industrial dump operators are not integrated among themselves or as part of the salmon production chain (Ingenieria_Alemana_S.A. 2006).

On the other hand, the sector does not receive direct investment from the State. It is a private sector issue. In fact, for example, in the Los Lagos Region, the industrial sector is made up of two large companies, while the rest are small operators. The latter existed before current environmental requirements were implemented (declarations and environmental impact studies). Therefore, many of them only have sanitary permits. The Superintendent of the Environment explains that they are also limited in terms of investment in machinery or technology to improve their processes. Paradoxically, those seeking to integrate ecological considerations consider the current environmental requirements that apply to this type of project as barriers. Since the rule is not retroactive, it would maintain bad practices in older operators. As indicated by an industrial landfill operator: "*the issue is that all the landfills going to the island, all the landfills that are operating on the island, are 30-year-old regulations. And there is nothing they can do. So of course, you see plans to close some landfills, which are a mess. Others, when you go to the operation, they are disgusting, they do not comply with anything. But they take refuge in the fact that "hey, they approved this 30 years ago and this is all they asked of me, I am complying". So, as the regulations*

[3] According to Moore (2014) cheap nature is the availability of raw materials as a strategy to advance capitalism through the construction of nature as external to human activity, making it easily accessible to firms for extraction of value. The speed in which global resources were incorporated into capitalist production greatly transformed ecological interactions affecting the capacity of ecosystems to react and sustain the current mode of production (hence, end of cheap nature) but also reconfigured understanding value of labor and commodities.

are not retroactive (...). The regulations today are demanding, and the communities are demanding, and you must comply with that and that is fine, that is very good. The issue is what happens with all the others that are behind".

Given the lack of industrial waste collection sites, an alternative is to transfer waste. However, it increases production costs, which rise with distance and the amount of waste. A relevant actor in the salmon industry points out that salmon farming can afford this cost, but other smaller-scale activities cannot—for example, Mitiliculture[4]: "*(...) for us as salmon producers, still the cost and given what the industry produces, still allows us to pay that cost, but there is an industry that does not, for example, the fish farmers, who in general are small fishermen who have converted to small-scale aquaculture... they are very complicated because they... [...] the costs do not allow them to make that transfer*".

Furthermore, the transfer of waste, in the context of the distances involved, is dangerous and compromises the proximity principle (Stack Lara et al. 2018). According to the Pollutant Release and Transfer Register between 2013 and 2018, other regions received about 25% of industrial waste in the Los Lagos Region. The main destination: Metropolitan Region of Santiago. In Aysén, close to 70% of industrial waste went to the Los Lagos Region; and in Magallanes, only 10% of the industrial waste left the Region. The numbers are eye-catching, considering the limited capacity of the Los Lagos Region to receive waste, and that Magallanes only has transfer stations (they only offer temporary storage for later use in other regions).

9.5 Waste as Opportunities

While for some producers, the waste issue has become an obstacle; for others, it has been an incentive to create businesses associated with valorisation. In this sense, it is interesting to note the points raised by Foley and Havice {Foley, 2016 #1782} as to the power dynamics and effects on accumulation that such schemes produces, while Le Billon and Spiegen (2021) {Le Billon, 2021 #1783} criticizes this so-called solutions as hiding new social harms and territorial inequalities.

The Clean Production Agreements (CPA), for example, have functioned as an opportunity for association and investment on the part of the State toward the salmon sector (Cárdenas 2016). The first APL was developed between 2003 and 2004, increasing investment in waste management (USD 91 million), reducing liquid industrial waste and increasing recycling (SalmónChile 2008). Although the private sector recognizes the CPAs positively, the initiative is limited to the time of investment of the program. Moreover, motivations are not purely environmental, as the CPAs offers individual benefits to the companies, favoring them at the time of certification.

[4] Miticulture is the artisanal cultivation of species of the family Mitilidae (mussels) (AMI Chile 2019).

9.6 The CE as a Solution to the Growing Waste Problems

To solve the dilemma of unavailability of industrial landfills and increasing waste generation, the transition toward a CE is increasingly called for by Chilean policymakers and the salmon industry. The Chilean government considers the CE as a strategy to reach its sustainability goals (Government_of_Chile 2020). In 2018 it established a Circular Economy Office under the ministry of environment and is currently in the process of elaborating a roadmap together with different industries, NGOs, and scientists to guide the economic transition. For the salmon industry, the CE represents both the solution to their waste problems, as well as a marketing tool.

On the one hand, circularity potentially reduces the amounts of waste. Thus, besides reducing environmental impact, waste disposal costs can be avoided and even revenue may be generated by valorising previous waste materials. On the other hand, international markets and consumers demand sustainable practices, and circularity in production systems improves the salmon companies' images, improving the marketing of their product. Thereby, it is of secondary importance if the absolute amount of waste is in fact reduced. Circularity, even if only applied on a small and insignificant scale, can serve as a marketing tool to better sell to far distant international customers.

Our research shows that the CE in the salmon industry is currently in its beginnings. Existing projects are of small scale and do not yet result in significant changes in production, environmental impact, or regional development. Companies with circular business models that can offer services to the salmon industry are few. An official of a medium-sized salmon company reported that "*in the concept of the CE, we are very far from having accomplished something. It is a topic that is just recently being conversed. No one understands what it is about. [...] The work, the efforts, and the advances that are being made are a little light.*" Similar affirmations have been issued by various interview partners, also from CE related businesses.

The industry's CE efforts are predominantly related to recycling projects, a practice already established before the CE gained momentum. Although a recycling rate of 24.36% between 2011 and 2015 demonstrates that recycling rates can be improved (Stack Lara et al. 2018), little focus is set on an overall reduction or increased reuse of materials, the key practices of a CE (Stahel 2014). The existing recycling practices have improved the industry's efficiency in waste management, but do not constitute a systemic change from a linear economy to a CE. Figure 9.3 indicates some of the CE operators for the salmon industry at present. Although the number of companies engaged in circular activities, mainly through recycling, as well as the quantities of waste processed by these companies has increased over the last years, their absolute impact in comparison with the total amounts of waste is still marginal, as affirmed by several interviewed companies. Besides the circumstance that the movement toward more circularity is still young and companies have to develop over the following years, the small impact of the CE today also stems from a variety of challenges

Table 9.2 Principal operators in CE in association with the salmon industry. Own elaboration

Operator	Year creation	Input material	Material produced
BECO	2019	Polyesterene	Sustainable concrete
Atando Cabos		Plastic stranded from the salmon industry at beaches	Ropes
Food 4 Future (F4F)	2015	Organic waste and insects	Animal food and fertilizers
Greenspot	2017	Plastic, polystyrene foam and industrial waste	Different kinds of plastics (trash cans, nets, water racks, compost bins)
Plásticos Puelche	2006	Pipes from the salmon industry	New pipes
Recollect	2015	Plastic, polystyrene foam and nets, and cables	Pallets, plastic boxes, polystyrene

arising in its implementation. Some of these challenges are typical for CE implementations, whereas others, as will be discussed in the following, are subject to the specific context, salmon aquaculture in a resource periphery (Table 9.2).

9.7 Challenges: Value Chain, Production Systems, and Chilean Policy

The establishment of circularity in the salmon industry is hampered by the materiality salmon itself and the corresponding value chain. Salmon aquaculture farms are in open waters, which make the materials used especially vulnerable to natural deterioration processes. Wastes are dissipated underwater, are contaminated by organic materials, and must be recollected, separated, and cleaned in work-intensive processes before they can be reused or recycled. "*It is precisely these processes that are most expensive in the value chain of recycling*", as stated by the CEO of a recycling firm. Therefore, large quantities of wastes are not recycled. In addition, some wastes, such as faeces or left-over feed, cannot be recollected and remain on the bottoms of Patagonian channels and fiords (Quinones et al. 2019).

A particularly complex matter for a CE in salmon farming is the fish mortalities. Being one of the most present wastes in terms of quantities and social impact—the inadequate disposal of 4,600 tons of dead salmon in 2016 had contributed to an algal bloom and resulted in protests and blockades of roads for weeks (Franklin 2016; Armijo et al. 2020)—it is also one of the wastes that has been treaded in a circular way even before the CE became a topic in the salmon industry. Under "normal" conditions, meaning a regular production with stable mortality rates, dead fish are largely processed into silage and re-enter the production cycle as fishmeal or are used for other agricultural purposes. However, this circular mode of waste treatment fails in times of massive salmon mortality, as has been the case during the 2016 algal bloom. When confronted with unforeseen quantities of dead salmon, the capacities of ensilage and other treatment facilities are surpassed, and the circular treatment of the mortalities is interrupted. They must be disposed of as waste, and given the complicated situation in the regional landfills, become a major problem, as illustrated by the 2016 incident.

Furthermore, in order to contribute to a sustainable development through emissions reductions, avoidance of material losses and strengthening of local economies, a CE has to be predominantly a regional activity (Stahel 2014). However, the Chilean salmon industry, as a resource periphery, represents the opposite. The salmon value chain is long, parts of fish-feed are imported from foreign countries, and salmons are produced in Chile and exported to international markets in North America, Asia, and Europe. An interview partner from the recycling industry affirmed that "*large quantities of recycled materials are sent to distant regions or other countries for processing and sale*", due to the absence of required processing infrastructure and a local market. Hence, it remains questionable as to what extend resource flows can be closed in an industry located in a resource periphery that extracts salmon which is then consumed on other continents.

Also, a CE involves increased cooperation between actors along the value chain to enable the necessary exchange of materials and information (de Man 2016; Winans et al. 2017). For CE in Chilean salmon aquaculture, the need for cooperation is exacerbated by the complex and long value chain of the industry. Moreover, the existing business culture in Chile, characterized by strong competition and individualism, amplifies the cooperation barrier. Potential material and knowledge exchanges are not conducted because companies work in an atomized way to achieve the highest profits for their stakeholders. A government official affirmed that "*there are great opportunities in changing the structures of value chains, which include many actors, but as the culture to cooperate does not exist, this becomes more difficult.*"

Similar situations have occurred in other CE-related contexts, for instance, in China, where policymakers have been identified as responsible for acting as coordinators and stimulating cooperation (Naustdalslid 2014; Winans et al. 2017). The recent efforts toward a CE made by the Chilean government have repeatedly been criticized in our interviews, and the government has been accused of failing its responsibility in leading the CE transition. For instance, a local entrepreneur engaged in the CE acknowledged that "*the government has been doing important work in promoting circular projects. But they are not enough. Most of the existing projects exist because people use their own resources*". Besides the lack of project funding and stakeholder coordination, CE supportive policies, such as more restrictive environmental and waste legislation, and knowledge provision, are absent or do not sufficiently stimulate the CE transition. The existing regulatory framework does not incentivize the avoidance or the reutilization of wastes, a factor considered as crucial for the success of a CE (Andersen 2006; Govindan and Hasanagic 2018; Li and Yu 2011). The 2016 Law EPR does not yet apply to the salmon industry. Furthermore, existing regulations hamper the CE. For instance, the reutilization of aquaculture sludges for agricultural purposes, a common practice in Norway and Scotland (Stack Lara et al. 2018), is prevented by existing sanitary legislation that has not yet been updated.

It is hoped by both state authorities and the salmon industry that the CE Roadmap will contribute to a more determined CE policy concerning legislation, as well as funding, coordinative, and knowledge support.

9.8 Conclusions

As mentioned, the CE in the Chilean salmon industry is still in an early stage and faces a variety of challenges that have to be overcome. The necessary changes in policy, production systems, and business culture will require time to be conducted. However, the ecological pressure from salmon farming is already high, waste treatment facilities', and landfills' capacities are exceeded. In other words, there is no CE in the Chilean salmon industry because there is not a system in place to make it work. In fact, the actual CE is a lot like the waste management in the salmon regions: lack of State presence, small integration between salmon firms and CE operators, and little participation from communities in the process. As such, waste treatment has become the new resource frontier associated with the salmon project in Chile. This represents an opportunity to expand and consolidate the salmon frontier, through the emergence of new actors and state-led policies to ease the creation of this new market, there is also a sense of urgency for solutions to concerned consumers in global markets. In other words, depending on the technological solutions created and human capital attracted to this new path, the region could remain an extractive-export-oriented economy or move into a service-high tech industry region. As it is, CE makes sense to firms because it creates value. However, we identify the following scalar implications that the existing challenges for implementing CE pose to the salmon industry:

Politically, the industry and the state need to be more ambitious than a "path route" and expand to define participative governance mechanisms that actually incorporate local communities into the decision-making regarding waste. The current centralized administrative system runs against an effective decision-making, which results in narrowing the scope of action for firms to waste treatment as opposed to transforming the salmon industry in a value-adding territorial figure, which would put the region in the path of circularity. The constitutional moment that Chile started in 2019 and solidified in 2021 with the election of 155 constituents represents an opportunity to redefine territorial autonomy, local capture of revenues, and decision-making mechanisms, especially considering that an important number of the elected constituents came from social-environmental and territorial movements. At a global scale, the pressure that consumers groups is making over firms to make their practices transparent is also a factor to consider in defining the political governance of waste and salmon firms in general.

Economically, the challenge is to create a business environment that fosters innovation and emergence of new actors that put content and value into the CE of the salmon industry, expanding territorial development. It is also important that the scientific institutions give support (financial, institutional) to the enterprises already doing CE at a small scale and promote their expansion further away from Los Lagos Region, to Aysén or Magallanes. It is fundamental to facilitate proximity and articulation across administrative and institutional boundaries. As now, while globally, waste has become the new frontier for accumulation, the lack of actors sizing the moment in the regions is a reaction to the lack of state support for I&D.

As such, the available time to realize a systemic change to a sustainable CE is limited. A broader understanding of circularity that moves away from its focus on waste and understand it as the end of unlimited extraction of resources, along with the articulation with diverse economies, could have a stronger impact on the sustainability of the industry in the long term. We believe that a shift must be accelerated by more political pressure from the state to the salmon companies to become more circular. However, considering the history of the industry, controls, and monitoring systems are necessary to ensure proper implementation.

A central question that arises during the implementation of a CE is if it shall indeed represent an alternative economic system that changes the dominant paradigm of resource extraction at a limited benefit for the local population and the cost of environmental degradation. Alternatively, the CE in the Chilean salmon industry may end up being just another form of an unsustainable extractive economy following a linear mode of production with circular elements. For the former to become a reality, the highlighted challenges have to be overcome and, most importantly, the existing business culture and economic goals have to change. Efficiency improvements through circularity will not benefit the environment if growth rates continue to be high, and waste generation is not reduced in absolute, but only in relative terms. Hence, absolute environmental limits, such as the carrying capacities of marine and terrestrial ecosystems, have to be incorporated in CE goals, and respected and protected by policymakers, the industry, and the civil society. If not, instead of the establishment of a sustainable CE, efforts will ultimately relate to more efficient waste management of an unsustainable industry. The statement "*you cannot punish yourself for producing more wastes if you grew by double*", made by a salmon company representative, exemplifies the need for a paradigmatic change.

References

Alfaro Monsalve K (2015) Procesos de transformación, persistencia y semi-proletarización de las familias campesinas de Quellón, Chiloé (1990–2015). Universidad Austral de Chile

AmiChile (2019) MITILICULTURA. AmiChile. https://www.amichile.com/post/mitilicultura

Andersen MS (2006) An introductory note on the environmental economics of the circular economy. Sustain Sci 2(1):133–140

Armijo J, Oerder V, Auger P-A, Bravo A, Molina E (2020) The 2016 red tide crisis in Southern Chile: possible influence of the mass oceanic dumping of dead salmons. Marine Pollution Bulletin 150:110603

Barton J, Pozo R, Roman A, Salazar A (2013) Urban restructuring of globalized territories: a charaterization of the organic growth of the cities of Chiloe, 1979–2008. Revista de Geografía Norte Grande (56):121–142. <Go to ISI>://WOS:000332749200007

Barton JR (1997) ¿Revolución Azul? El impacto regional de Ia acuicultura del salmón en Chile. Eure-Revista Latinoamericana De Estudios Urbano Regionales XXII(68):57–76

Barton JR, Gwynne RN, Murray WE (2008) Transformations in resource peripheries: an analysis of the Chilean experience. Area 40(1):24–33. <Go to ISI>://WOS:000254413100003

Bridge G (2008) Global production networks and the extractive sector: governing resource-based development. J Econ Geogr 8(3):389–419

Buschmann A (2001) Impacto ambiental de la acuicultura: el estado de la investigacion en Chile y el mundo. Un analisis bibliografico de los avances y restricciones para una produccion sustentable en los sistemas acuaticos

Buschmann A, Farías L, Tapia F, Varela D, Vásquez M (2016) Informe Final Comisión Marea Roja

Bustos-Gallardo B, Irarrazaval F (2016) Throwing money into the sea: capitalism as a world-ecological system. Evidence from the chilean salmon industry crisis, 2008. Capital Nat Social. https://doi.org/DOI:10.1080/10455752.2016.1162822

Bustos B, Román Á (2019) A sea uprooted: islandness and political identity on Chiloé Island, Chile. Island Stud J 14(2):97–114

Bustos-Gallardo B (2012) Brote del virus ISA: crisis ambiental y capacidad de la institucionalidad ambiental para manejar el conflicto. Eure-Revista Latinoamericana De Estudios Urbano Regionales 38(115):219–246

Bustos-Gallardo B (2013) The ISA crisis in Los Lagos Chile: a failure of neoliberal environmental governance? Geoforum 48(0):196–206. https://doi.org/10.1016/j.geoforum.2013.04.025

Bustos-Gallardo B, Delano J, Prieto M (2019) "Chilote tipo salmón"- relaciones entre comodificación de la naturaleza y procesos de producción identitaria, el caso de la región de los lagos y la industria salmonera. Estudios Atacamenos

Cabello FC (2004) Antibióticos y acuicultura en Chile: consecuencias para la salud humana y animal. Rev Med Chil 132(8):1001–1006

Cárdenas JC (2016) El fraudulento acuerdo salmonero de "producción limpia.". El Desconcierto (June 16th). https://www.eldesconcierto.cl/2016/06/16/el-fraudulento-acuerdo-salmonero-de-produccion-limpia/

Cid Aguayo BE, Barriga J (2016) Behind certification and regulatory processes: contributions to a political history of the Chilean salmon farming. Glob Environ Change 39:81–90

de Man R, Friege H (2016) Circular economy: European policy on shaky ground. Waste Manag Res 34(2):93–95

Dorn FM, Huber C (2020) Global production networks and natural resource extraction: adding a political ecology perspective. Geographica Helvetica 75(2):183–193

Fløysand A, Barton JR, Román Á (2010) La doble jerarquía del desarrollo económico y gobierno local en Chile: El caso de la salmonicultura y los municipios chilotes. EURE (Santiago) 36:123–148. http://www.scielo.cl/scielo.php?script=sci_arttext&pid=S0250-71612010000200006&nrm=iso

Fortt A, Cabello F, Buschmann A (2007) Residuos de tetraciclina y quinolonas en peces silvestres en una zona costera donde se desarrolla la acuicultura del salmón en Chile. Rev Chilena Infectol 24(1):14–18

Franklin J (2016) Toxic "red tide" in Chile prompts investigation of salmon farming. Retrieved May 5, 2020, from The Guardian website: https://www.theguardian.com/world/2016/may/17/chile-red-tide-salmon-farming-neurotoxin

Gallo GS (2001) El Rostro de una Nueva Identidad: La Expansión de la Industria Salmonera en el Archipiélago de Los Chonos IV Congreso Chileno de Antropología. Colegio de Antropólogos de Chile A.G., Santiago de Chile

Garcés J (2018) 2018: Exportaciones de salmón chileno anotan US$ 5.157 millones. SalmonExpert. https://www.salmonexpert.cl/article/2018-exportaciones-de-salmn-chileno-anotan-us-5157-millones/

Gargallo E, Kalvelage L (2020) Integrating social-ecological systems and global production networks: local effects of trophy hunting in Namibian conservancies. Dev South Africa 1–17

Geissdoerfer M, Savaget P, Bocken NMP, Hultink EJ (2017) The circular economy—a new sustainability paradigm? J Clean Prod (143):757–768. https://doi.org/10.1016/J.JCLEPRO.2016.12.048

Government_of_Chile (2020) Chile's Nationally Determined Contribution: Update 2020. https://www4.unfccc.int/sites/ndcstaging/PublishedDocuments/ChileFirst/Chile%27s_NDC_2020_english.pdf
Govindan K, Hasanagic M (2018) A systematic review on drivers, barriers, and practices towards circular economy: a supply chain perspective. Int J Prod Res 56(1–2):278–311
Hayter R, Barnes TJ, Bradshaw MJ (2003) Relocating resource peripheries to the core of economic geography's theorizing: rationale and agenda. Area 35(1):15–23
InduAmbiente (2012) Futuro Naranja. Los beneficios del ensilaje para manejar la mortalidad de los salmones, pp 128–129
INE (2018) Industria salmonera tuvo ingresos por más de 5 mil millones de dólares y generó 21.462 puestos de trabajo en promedio en 2016. https://www.ine.cl/prensa/2019/09/16/industria-salmonera-tuvo-ingresos-por-m%C3%A1s-de-5-mil-millones-de-d%C3%B3lares-y-gener%C3%B3-21.462-puestos-de-trabajo-en-promedio-en-2016
Ingenieria_Alemana_S.A. (2006) Estudio diagnóstico en la generación y gestión de los residuos que genera la actividad productiva de determinados sectores económicos regionales, principalmente residuos industriales sólidos [Informe diagnóstico]
Irarrazabal F, Bustos B (2018) Global Salmon networks: unpacking ecological contradictions at the production stage. Econ Geogr. https://doi.org/10.1080/00130095.2018.1506700
Jacobsen JA, Hansen LP (2001) Feeding habits of wild and escaped farmed Atlantic salmon, Salmo salar L., in the Northeast Atlantic. ICES J Mar Sci: J Conseil 58(4):916–933
Kirchherr J, Reike D, Hekkert M (2017) Conceptualizing the circular economy: an analysis of 114 definitions. Resour Conserv Recycl 127:221–232. https://doi.org/10.1016/J.RESCONREC.2017.09.005
Korhonen J, Honkasalo A, Seppälä J (2018) Circular economy: the concept and its limitations. Ecol Econ 143(37–46)
Le Billon P, Spiegel S (2021) Cleaning mineral supply chains? Political economies of exploitation and hidden costs of technical fixes. Rev Int Political Eco 1–27. https://doi.org/10.1080/09692290.2021.1899959
Leon J, Tecklin D, Farias A, Diaz S (2007) Salmon farming in the Lakes of Southern Chile—Valdivian Ecoregion. History, tendencies and environmental impacts
Leon Munoz J, Soto D, Fuentes M, Montes RM, Quinones RA (2019) Environmental issues in Chilean salmon farming: a review. Rev Aquacult 11(2):375–402. https://doi.org/10.1111/raq.12337
Li J, Yu K (2011) A study on legislative and policy tools for promoting the circular economic model for waste management in China. J Mater Cycles Waste Manag 13(2)
Luxoro C (2018) "Historia del Huirihue en Chile" Florecimientos Algales Nocivos (APP, Issue)
Mansilla Torres S (2006) Chiloé y los dilemas de su identidad cultural ante el modelo neoliberal chileno: La visión de los artistas e intelectuales. Alpha (Osorno) 9–36. http://www.scielo.cl/scielo.php?script=sci_arttext&pid=S0718-22012006000200002&nrm=iso
Mascareño A (2018) Controversies in social-ecological systems: lessons from a major red tide crisis on Chiloe Island, Chile. Ecol Soc 23(4):15
Millanao A, Gómez C, Tomova A, Buschmann A, Dölz H, Cabello FC (2011) Uso inadecuado y excesivo de antibióticos: Salud pública y salmonicultura en Chile. Rev Med Chil 139(1):107–118
Miller JC (2018) No fish, no mall. Industrial fish produce new subjectivities in Southern Chile. Geoforum 92:125–133
Moore JW (2015) Capitalism in the web of life: ecology and the accumulation of capital. Verso Books
Mowi (2020) Salmon farming industry handbook 2020
Naustdalslid J (2014) Circular economy in China—the environmental dimension of the harmonious society. Int J Sust Dev World 21(4):303–313
Neira Flores Á (2018) Chiloé, mayo del 2016. El mayo chilote y los discursos desplegados en el conflicto socioambiental del archipiélago de Chiloé

Planells Gutiérrez D (2019) El nuevo Quinchao: trabajo asalariado, migración y lazo social en el Chiloé contemporáneo. Academia de Humanismo Cristiano], Santiago

Quinones RA, Fuentes M, Montes RM, Soto D, León-Muñoz J (2019) Environmental issues in Chilean salmon farming: a review. Rev Aquac 11(2):375–402

Rammelt C, Crisp P (2014) A systems and thermodynamics perspective on technology in the circular economy. Real-World Econ Rev 68:25–40

Román Á, Barton JR, Bustos B, Salazar A (eds) (2015) Revolución salmonera: paradojas y transformaciones territoriales en Chiloé. RIL Editores

SalmónChile (2008) Salmonicultura y sustentabilidad. https://lyd.org/wp-content/uploads/2011/06/lionel-sierralta-marzo2008.pdf

SalmonChile (2020) Cifras de empleo. Salmon Chile. https://www.salmonchile.cl/salmon-de-chile/cifras-de-empleo-salmonchile/. Accessed 20 Oct 2020

Schroeder P, Anggraeni K, Weber U (2019) The relevance of circular economy practices to the sustainable development goals. J Ind Ecol 23(1):77–95

SERNAPESCA (2020) Escape de peces en la salmonicultura 2010–2020

Stack Lara I, Aranda Cuadro MC, Casas-Cordero E (2018) Informe Final Proyecto FIPA N ° 2016–69. ID: 4728–103-LE16 "Establecimiento de las Condiciones Necesarias para el Tratamiento y Disposición de Desechos generados por Actividades de Aquicultura"

Stahel W (2014) Reuse is the key to the circular economy | Eco-innovation Action Plan. https://ec.europa.eu/environment/ecoap/about-eco-innovation/experts-interviews/reuse-is-the-key-to-the-circular-economy_en. Accessed 22 March 2020

Subpesca (2017) Informe Sectorial de Pesca y Acuicultura 2017. http://www.subpesca.cl/portal/618/articles-95982_documento.pdf

Superintendencia_del_medioambiente (2020) Estadísticas de Fiscalizaciones Región de Los Lagos. https://public.tableau.com/views/Fiscalizaciones-mobile/Fiscalizaciones-Mobile?:embed=y&:showVizHome=no&:host_url=https%3A%2F%2Fpublic.tableau.com%2F&:tabs=no&:toolbar=yes&:animate_transition=yes&:display_static_image=no&:display_spinner=no&:display_overlay=yes&:display_count=yes&:loadOrderID=0

Swanson HA (2015) Shadow ecologies of conservation: co-production of salmon landscapes in Hokkaido, Japan, and southern Chile. Geoforum 61:101–110. https://doi.org/10.1016/j.geoforum.2015.02.018

Terram (2018) Antecedentes Económicos de la Industria Salmonera en Chile (CARTILLA INFORMATIVA N° 2, Issue)

Terram F (2019) El régimen jurídico-ambiental de la salmonicultura en Chile (Cartilla Informativa N°1, Issue). https://www.terram.cl/descargar/recursos_naturales/salmonicultura/cartilla/El-regimen-juridico-ambiental-de-la-salmonicultura-en-Chile.pdf

Undersecretary_of_Regional_Development (2018) Diagnóstico de la situación por comuna y por región en materia de RSD y asimilables (Diagnóstico y Catastro de RSD Año 2017). http://www.subdere.gov.cl/sites/default/files/documentos/2_marco_legal_agosto_2018.pdf

Winans K, Kendall A, Deng H (2017) The history and current applications of the circular economy concept. Renew Sustain Energy Rev 68:825–833. https://doi.org/10.1016/j.rser.2016.09.123

Zink T, Geyer R (2017) Circular economy rebound. J Ind Ecol 21(3):593–602

Beatriz Bustos is an associate professor at the Department of Geography, University of Chile. She holds a Ph.D. in Geography and a Master in Public Administration from Syracuse University, a Master in Anthropology and Development and a BA in Public Administration from Universidad de Chile. Her research focuses on resources geography and the sociopolitical transformations that exploitation of natural resources produces in rural communities. From a political ecology perspective, she has published about the production of commodity regions, the effects of salmon and mining economies in rural landscapes and currently leads a research on extractive citizenship sponsored by an ANID grant.

María Inés Ramírez obtained degrees in Geography (B.Sc.) at University of Chile and M.Sc. in Geography at the same university. After working in investigation at sustainable urban development and salmon aquaculture in Chile, she works now as international consultant on local sustainability in southern Chile.

Marco Rudolf obtained degrees in Mathematics (B.Sc.) and Governance of Risks and Resources (M.Sc.) at Heidelberg University. After investigations in circular economy and salmon aquaculture in Chile, he has worked on hydrogen and its future role in the German energy system.

Chapter 10
No worker's Land. The Decline of Labour Embeddedness in Resource Peripheries

Miguel Atienza

> "... unlike other commodities, labour power has to go home every night"
> (Harvey, The urban experience 1989, 19)

Abstract Local development literature has generally assumed that labour force is a crucial and territorially embedded asset, the origin of social and political demands to improve the quality of life and one of the main driving forces of local change and economic development. Nowadays, this assumption is far from true, particularly in the case of resource peripheries. As it has happened with the increasing externalization of tasks to service firms located in the main urban agglomerations, a growing proportion of workers—especially those in primary industries—live far away from the resource peripheries where they produce, fostered by the increasing use of Fly-in/fly-out (FIFO) and other systems and technologies facilitating long-distance work. Furthermore, the increasing use of robotics and remote work in primary activities can turn resource peripheries into places for production, but not for social reproduction. By using some of the main elements of GPN framework, this chapter proposes an extension of resource periphery research agenda, considering this trend of declining labour territorial and social embeddedness; analyses how this trend has become a distinctive characteristic of resource peripheries; and what the consequences are for economic development in these territories.

Keywords Resource periphery · Long-distance commuting · Natural resources · Regional development

10.1 Introduction

Today, the research agenda on resource peripheries proposed by Hayter et al. (2003) is more urgent than ever, due to the significant increase in works that cast doubts

M. Atienza (✉)
Departamento de Economía, Instituto de Economía Aplicada Regional (IDEAR), Universidad Católica del Norte, Antofagasta, Chile
e-mail: miatien@ucn.cl

F. Irarrazaval and M. Arias-Loyola (eds.), *Resource Peripheries in the Global Economy*, Economic Geography, https://doi.org/10.1007/978-3-030-84606-0_10

on the opportunities for natural resource-led regional development during the last two decades, particularly after the end of the commodity prices super-cycle in 2011 (Phelps et al. 2015; Scholvin et al. 2020). Furthermore, the growing relevance of natural resources in Global Production Networks (GPN) in the context of the Anthropocene makes resource peripheries essential territories for achieving a sustainable form of global development (Barton et al. 2008; McElroy, 2018).

This chapter aims to extend the concept and agenda of resource peripheries in two senses: first, reformulating the way sociospatial relations are predominantly conceived in resource peripheries analysis. This, by going from a conception mainly based on space, understood as "territory" and "place", to another where resource peripheries are analysed in terms of sociospatial relations of "scale" and "networks", according to the classification proposed by Jessop et al. (2008). The second extension is related to a deeper analysis of the role of labour in the economic development of resource peripheries. Traditionally, labour has been conceived as a deeply rooted local asset, the origin of social movements, contestation and demands for improving workers´ rights, the quality of life of local communities and the supply of public goods in resource peripheries. Such view is being currently challenged, particularly in the case of natural resource-intensive industries, due to the increasing use of automation, long-distance commuting and remote work (Aroca and Atienza 2011; Manky 2016, 2018; Paredes and Fleming-Muñoz 2021). Unlike Harvey's quote mentioned at the beginning of this chapter, today labour power does not need to go home every night and, in the next future, labour power might not need to go to the production sites, especially in some activities and occupations. This chapter defends that these changes in the organization of production lead to a decline in the social and territorial embeddedness of workers in resource peripheries and, subsequently, limit their development opportunities. For this purpose, some of the elements of GPN framework are instrumentally used to complement both extensions of the concept of resource periphery, and to understand where value is captured when workers become socially and territorially disembedded and become an extra-local asset.

It is important to acknowledge that the increase in automation, long-distance commuting and remote work is a global trend, but it does not affect equally all primary activities. In particular, it is more spread in gas and mining than in other industries. For this reason, the majority of the references used in this chapter study gas and mining extraction. It is plausible to think, however, that the consequences of the analysis presented here can be extended to resource peripheries in general and particularly to those located in remote and isolated places and where human settlements are relatively small. Furthermore, these trends in the organization of labour will tend to increase in the next future.

The article is divided into three parts: first, we propose how the agenda on resource peripheries can be re-examined considering the role of labour in a context of increasing use of automation, long-distance commuting and remote work that directly affects their position in national and global networks. The second section adopts a network approach to analyse the consequences of these changes for the development opportunities of resource peripheries using the GPN framework. Finally, the implications for the future research agenda on resource peripheries are discussed.

10.2 The Position of Labour in Resource Peripheries Agenda

The research agenda on resource peripheries proposed by Hayter et al. (2003) considers four institutional dimensions: industrialism (economic dimension), environmentalism (environmental dimension), aboriginalism (cultural dimension) and imperialism (geopolitical dimension) and described these areas mainly as contested and conflict spaces essential to understand globalization whose analysis has been mostly focussed on the core regions. Surprisingly, one missing piece in this initial proposal was the labour force, not only in terms of the role of workers as a crucial local asset and agent of development, but also in terms of how labour is singularly organized in this type of territory. More recent works briefly point out labour control and conflicts around resource extraction (Barton et al. 2008) as one of the characteristics and part of the contested nature of resource peripheries. In most cases, however, labour force remains almost absent of the analysis (Argent 2017; Breul and Revilla Diez 2018; McElroy 2018), except in some works that, from a Marxist perspective, pay special attention to the role and organization of labour at a global scale (Arboleda 2020).

This absence in the analysis of resource peripheries is striking, especially considering that many of the extractive activities have been, to some extent, pioneers of what Baldwin (2019) called the "third wave of globalization". According to this authors, the reduction in the trade costs of goods and ideas characterized the first and second waves of globalization, respectively. In contrast, the "third wave of globalisation" that is now starting, directly affects the mobility of workers in different ways: (1) they are substituted by machines through automation; (2) they do not need to stay in the production sites, thanks to remote work; and (3) workers are able to separate their sites of production from their sites of social reproduction due to technological and institutional changes that allow shift work. These trends are not new in natural resource-intensive activities, which have experienced a persistent, a continuous reduction in labour intensity, have developed increasing technologies for remote work and were one of the first industries in implementing shift work systems that allow different forms of long-distance commuting (Arboleda 2020). Indeed, these are general trends that, in the long term, will permeate most industries, but are especially relevant in the case of peripheral regions because: the incentives for workers to live in peripheral and remote areas are low; these trends will increase in the next future; and the direct consequences that these changes in the organization of labour will have on the development opportunities of resource peripheries.

Labour is a strategic local asset for regional economic and social development. It is generally agreed that a thick and skilled local labour market is essential for increasing productivity, attracting more competitive companies and for the subsequent integration of regions into global networks. Furthermore, labour skills are a crucial factor for upgrading in value chains and for the creation of new path and diversification through innovation. Additionally, workers' organizations, such as unions, strengthen local institutions, which makes them a relevant agent for mediation and arbitration that would allow value capture in resource peripheries. From these perspectives, labour has been traditionally considered as a socially and territorially embedded local agent. Currently, however, this assumption is at least controversial and particularly in the case of resource peripheries.

The fragmentation of production and the creation of spatial divisions of labour since the 1970s (Massey 1984) are increasingly extended to the labour markets. The geography of work has been progressively fragmented, giving rise to new forms of organization that separate workers from the communities of their sites of production and limit their role as local development agents (Manky 2016). In this sense, it is necessary to incorporate this decline in the embeddedness of workers from local communities as an essential feature of the research agenda on resource peripheries, where shift work systems, long-distance commuting as well as remote work are widespread but have received scarce attention (Sheppard 2013; Vodden and Hall 2016; Martinus et al. 2020). While in this chapter it is contended that this trend is a relevant characteristic of labour in resource peripheries in general, it is important to acknowledge that its effects are not even across the variegated forms that resource peripheries can take. The decline in labour embeddedness is deeper in isolated and remote peripheries, where urban agglomerations are smaller and where local workers can be almost reduced to a minimum. This decline is also more spread in the mining regions than in territories specialized in agriculture, cattle, fishing and forestry.

The analysis of resource peripheries just as places or territories is insufficient due to the increasing fragmentation of production and labour markets. In this sense, it is crucial to adopt a scale or a network conception of sociospatial relations (Jessop et al. 2008) for a better understanding of how the multiscalar organization of production and labour determines the development opportunities of resource peripheries. In fact, many workers in resource peripheries are not a local asset anymore, and belong to a network of urban centres that forms a hierarchy of places, at a national and even transnational scale, directly affecting the contested nature of resource peripheries.

This type of organization of production combines the interests of both the lead firms through strategies of spatial divisions of labour (Massey 1984); and of some groups of workers that decide to separate production and social reproduction sites, to benefit from higher real wages and, at the same time, be close to their relatives or to urban amenities that are not available at resource peripheries (Manky 2016; Paredes et al. 2018). While there is not a complete agreement about to what extent both firms and workers obtain the same benefits[1] yet, it seems clear that these type of organizations of spatial labour relations go to the detriment of resource peripheries development opportunities, as it will be argued below.

[1] While most workers tend to prefer shift work systems and earn higher wages (Manky, 2016; Paredes, Soto and Fleming, 2018), many studies have pointed out the negative effects of long-distance commuting on health, family life, unions´ bargaining power and subjective wellbeing (Richalet et al., 2002; Tokington et al., 2011; Palomino and Sarrias, 2019).

10.3 The Decline of Labour Embeddedness and Value Capture in Resource Peripheries

The fragmentation of labour across space implies adopting a scale and network conception of sociospatial relations for understanding the consequences of automation, commuting and remote work in the development resource peripheries. For this purpose, some elements of the GPN framework[2] are instrumentally used in this section following previous contributions that use GPN for the analysis of extractive industries (Bridge 2008; Irarrazabal 2021).

Long-distance commuting, unlike traditional urban commuting, means that travelling to and from work on a daily basis is not necessary. Shift work established a fixed number of days on the job, followed by a fixed number of days at home (Hobart 1979). This allows most long-distance commuters to live in the main cities of the urban hierarchy (Aroca and Atienza 2011; Manky 2016; Martinus et al. 2020; Prada-Trigo et al. 2021). In the case of remote work, some functions of the production process can be performed without the need of travelling to the resource periphery or doing so only exceptionally. Remote workers can live elsewhere, but they tend to prefer the sites where companies' headquarters are located, usually the main cities of the country or even abroad. In both cases, this form of organization of labour introduces resource peripheries in a network of urban centres that compete for the location of workers' residence and disconnect workers from the local communities of resource peripheries. Finally, robotics and automation implies the substitution of workers that traditionally lived in resource peripheries (Paredes and Fleming-Muñoz 2021).

The three dimensions proposed by the GPN 1.0 framework—the network, social and territorial embeddedness; the distribution of power within the network; and the creation, enhancement and capture of value—are particularly useful to analyse the interaction between networks and territories as a way to understand how the disconnection of workers from the local communities at the sites of production affects development opportunities in resource peripheries.

10.3.1 Embeddedness

The most evident effect of automation, remote work and long-distance commuting is the decline in the territorial and social embeddedness of workers in the resource peripheries. Remote workers are only connected to these territories through a computer screen in integrated operation centres, while long-distance commuters live in camps or hotels detached from the rest of the local community. In this sense, such

[2] This chapter is fundamentally based on the so-called GPN 1.0 as a broad heuristic framework to understand uneven development. While some of the ideas of the GPN 2.0 (Coe and Yeung, 2015) are also taken into account, using this framework is beyond the scope of this chapter and can be part of future lines of research.

workers mostly coexist with the extractive communities in very limited periods of time: during their time at the workplace and while waiting for transportation to their homes (Garcés 2003; Manky 2016). The loss of a significant segment of workers locally installed also supposes a loss of social bonds between workers and with the local community, leading to a reduction in the social embeddedness of economic activity in the territory. Furthermore, the existence of potential conflicting interests between resource peripheries and non-local workers limits the role of labour as a local asset that creates and consolidates a local community able to fight for the improvement in the quality of life, which is considered one of the conditions to promote economic development (Coe 2012; Atienza et al. 2019).

This type of organization of labour not only affects the embeddedness of workers. At the same time, it reduces the dependence of extractive companies on local workers, which leads to weakening the territorial embeddedness of these companies. This is particularly relevant in the event of a crisis caused by either price-cycles or the potential loss of economic value of commodities, since the social effects on unemployment are widespread across the urban system, reducing potential demands and conflicts with local population and making disinvestment easier. Finally, remote work and long-distance commuting also reduce the network embeddedness of resource peripheries, which is their degree of functional and social connectivity within a GPN (Coe and Yeung 2015). In this sense, the use of workers—locally rooted—from the resource peripheries in the production of commodities is mainly limited to those functions that require spatial proximity, which, in most cases, are maintenance, cleaning and other routine tasks, reproducing some of the main features of the spatial divisions of labour described by Massey (1984). Furthermore, these occupations are among the most likely to be substituted by automation (Paredes and Fleming-Muñoz 2021).

10.3.2 Power

The distribution of power within the resource GPN is also affected by long-distance commuting and remote work, especially limiting the traditional role of workers—through labour unions—as local agents that mediate and arbitrate in the capture of value by resource peripheries. The fragmentation of local labour markets caused by both long-distance commuting and remote work can be used by leading firms for the application of "divide and rule" strategies, that tend to reduce labour workers' agency in a variety of ways. On the one hand, the decline of social and territorial embeddedness implies a potential atomization of labour demands, which might lead to individualistic negotiation strategies. On the other hand, the territorial fragmentation of workers can give rise to conflicting interests between them, which could lead to a negotiation of benefits for particular groups depending on their sites of social reproduction. This may leave aside potential demands for increasing the quality of life and other development conditions in resource peripheries.

Research analysing the way in which long-distance commuting affects the workers' agency in resource peripheries is still scarce. Within this literature, the studies by Manky (2016, 2018), based on the mining industry in Peru and Chile stand out. The author shows how long-distance commuting and the change to shift work systems negatively affected the capacity of collective action of mining workers and has significantly reduced the number of strikes both in Chile[3] and particularly in Peru, making it more difficult to carry out strong local mobilizations. At the same time, however, these changes fostered new organizational strategies by the unions, to maintain their bargaining power. While initially the implementation of shift work led to strikes and resistance on the part of mining workers, today it is generally accepted, and labour unions recognize spatial mobility as part of the workers' rights. This has led to bargaining strategies based on individual benefits, such as decentralized health systems, mobility bonus and on individual disputes with the mining companies, significantly increasing the number of particular lawsuits when workers' rights are violated.

Other bargaining strategies are related to the co-ordination, merging and association with other unions both at a national and transnational scale. At the same time, the characteristics of strikes have also changed. In the past, these protests had a strong support of dense communal networks, but today they are based on the strategic stoppage in some stages of the production process such as logistics. While some of these new strategies also include networking with regional groups and organizations concerned about the negative local externalities of extractive activity, it seems evident that the power of workers is not generally exerted to the benefit of the local community, and it predominantly tends to show a corporatism spirit. This leads to what Castree et al. (2004) called a "geographical dilemma": a situation where geographically fragmented labour agency can have negative knock-on effects in other places. In particular, this "geographical dilemma" negatively affects resource peripheries, whose bargaining power within the GPN significantly diminishes. Finally, the effects of remote work on power in resource peripheries have not been studied yet, but it is plausible to think that those effects will be even greater than in the case of long-distance commuting, due to the total disconnection of workers from the local communities where commodities are produced.

10.3.3 Value

In principle, value creation in resource peripheries—understood of the capacity of firms within the network of generating different forms of rent—does not seem to be affected by the relatively recent changes in the organization of labour. Furthermore, it would be reasonable to think that value added has grown due to the increasing labour productivity caused by automation and to reductions in the costs of production in fix investments and wages. However, value enhancement—considered as the possibility

[3] While the number of strikes in the Chilean mining industry has diminished in number and strength, it is still one of the activities where strikes are more common (Pérez, Medel and Veláquez, 2017), but it is debatable to what extent these demands are connected to local community demands.

of technology transfers and the improvement in the skill content of products, leading to industrial upgrading in resource peripheries—can be severely limited due to the scarcity of skilled workers and the existence of a local fabric mostly specialized in routine functions, hampering the development of local innovation capabilities, new path creation and productive diversification in the long term. Another plausible hypothesis is that the combination of the decline of labour social and territorial embeddedness and the reduction in the bargaining power of local organizations in resource peripheries is facilitating value capture by extra-local agents (both lead firms and non-local workers) to the detriment of resource peripheries.

The fragmentation of labour can be seen as a game with two winners (lead firms and non-local workers[4]) and a loser (the resource peripheries). Long-distance commuting and remote work allow lead firms to generate "human resources rents" and value capture by accessing a greater job supply (potentially global), which means paying lower wages, as well as a significant reduction in investment for the installation of workers. The latter, due to the passage from the expensive company town model to the relatively cheaper mining hotel and, more recently, to robotics and Integrated Operation Centres (Storey and Shrimpton 1988; Houghton 1993; Storey 2010; Paredes and Fleming-Muñoz 2021). Likewise, the reduction in the number of strikes and the possibility of implementing work journeys of 12 h, thanks to shift work systems, are other potential channels for firms to obtain "organizational rents" and value capture. Furthermore, as previously mentioned, the debilitation of local institutions reduces the bargaining power of the local community that, in many cases, is not sufficiently dense and organized to impose conditions for the application of local content and other policies oriented towards the development of resource peripheries, facilitating value capture by leading firms and reducing the opportunities for strategic coupling into GPN.

Since the GPN framework takes for granted that workers are an embedded local asset, it has not considered that, in resource peripheries, extra-local workers can also capture part of the value created. The first and most evident manner in which extra-local workers can capture value is through wages that are mostly expended in their sites of social reproduction. This loss of value, however, goes beyond the amount of money received by those workers, but it also includes the reduction in the multiplier effects of demand that, in the end, has indirect negative effects on the creation of new employment, on local entrepreneurship and on diversification (Aroca and Atienza 2011). Indirectly, the regions where extra-local workers live also capture "fiscal rents" that has been traditionally considered of local nature. In fact, long-distance commuting and remote work imply a reduction in the collection of tax revenues by local governments in resource peripheries, particularly in the property value taxes (Paredes and Fleming-Muñoz 2021).

[4] In this case, however, it is important not to forget, as previously mentioned, some negative effects on the health, family life and subjective wellbeing of workers.

10.3.4 Strategic Coupling and Regional Development

According to the analysis made in the previous sections, automation, long-distance commuting and remote work significantly affect the main dimensions of GPN 1.0 framework, and, in all cases, the consequences are negative for increasing the development opportunities of resource peripheries. According to the GPN framework, regional development is the outcome of the interaction between regional and extra-regional assets, actors and institutions and is effected by a process of "strategic coupling", where coupling implies a dynamic "fitting" between the local assets of regions, the leading and strategic firms of the network and their specific objectives, in a process which both transcends and interlinks territorial borders (Henderson et al. 2002; Coe et al. 2004; Coe and Yeung 2015).

On balance, the incorporation of changes in the organization of labour through automation, long-distance commuting and remote work suggests that coupling processes in resource peripheries become closer to the structural than to the indigenous or functional varieties defined by Coe and Yeung (2015). Unlike the indigenous type of coupling where autonomous local agents are able to reach outside the home region, to construct global networks and to capture the value locally created, the structural type of strategic coupling is defined by unequal power relations between lead firms within GPNs and host regions leading to a dependent type of development. These different coupling scenarios work themselves out not only at the international, but also at the national scale. This, by considering the position of resource peripheries in a hierarchy of places where the different sites of the national urban systems occupy distinctive positions in the spatial division of labour within GPNs, embedding economic activities and capturing disparate amounts of value.

The separation of production sites from social reproduction sites that is characteristic of the "third wave of globalization" reduces the role of labour as a local asset and an agent of development, debilitates local communities in resource peripheries and makes more likely a structural form of coupling leading to value capture by firms and non-local workers. While this structural type of strategic coupling has been more the rule than the exception in resource peripheries along history, this recent trend could increase their traditionally dependent position in GPN. Therefore, this situation, that has not been generally considered by resource periphery and GPN frameworks, tends to reinforce pre-existing core periphery patterns. In many cases, this situation can even lead resource peripheries to become new forms of enclave economies, one where most workers occupied in primary activities live outside of these territories (Arias et al. 2014; Phelps et al. 2015), facilitating the opportunities of decoupling and disinvestment by firms in the case of a contraction in prices or a loss of economic value of commodities.

10.4 Implications for an Extension of the Resource Peripheries Agenda

The analysis of the changes in labour organization through automation, long-distance commuting and remote work suggest the need of extending the research agenda on resource peripheries. Labour has been generally considered a local asset, capable of capturing value and promoting virtuous forms of strategic coupling through bargaining processes and contestation. While this role is still relevant and need to be analysed as part of the conflicting nature of resource peripheries, it is also important to consider what are the consequences that the new forms of organization of labour have and will have in a context of fragmented labour markets and where a significant part of the labour force can become extra-local due to long-distance commuting and remote work.

The first aspect to consider when incorporating the fragmentation of labour markets to the analysis of resource peripheries is the need to adopt a conception of sociospatial relations based on the ideas of scale and network, where the interaction among different spatial levels of analysis becomes crucial. In this chapter, a preliminary and instrumental use of the basic concepts of GPN 1.0 framework has been used, but future analyses need to be extended by considering the more dynamic a causal extensions proposed by GPN 2.0 framework, as well as other theoretical approaches such as those related to (dis)articulations (Bair and Werner 2011; McGrath 2018), sacrifice zones (Reinert 2018; Arboleda 2020) and enclave economies that show the dark side of economic geography and globalization (Arias et al. 2014; Phelps et al. 2018). Combining these lines of research is essential for a better understanding of both the contested nature of resource peripheries and the conditions that lead to uneven development across space.

The results presented could seem rather pessimistic about the development opportunities of resource peripheries. Future agenda, however, should not fall into determinism and try to discover what the strategies are that, in this new context, could lead to the value capture, diversification and development in resource peripheries. As in the case of unions' bargaining strategies adopted after the extension of long-distance commuting, new strategies for value capture need to be designed and implemented by local population in resource peripheries. These strategies can be thought both at a *place level*, such as the demand for local content policies, and at a *multiscalar level*, considering the association and networking with other regions, global NGOs and international organizations. In this sense, it is important to acknowledge that resource peripheries are diverse and that these strategies would change depending on the type of commodity and the size of human settlements established in each territory.

Acknowledgements This work has been supported by the FONDECYT Project number 1210765 "La fragmentación de la geografía del trabajo: La conmutación de larga distancia en la red de producción global de la minería en Chile" from ANID (Chile).

References

Arboleda M (2020) Planetary mining. Territories of extraction under late capitalism. Verso, London, UK

Argent N (2017) Rural geography I: resource peripheries and the creation of new global commodity chains. Prog Hum Geogr 41(6):803–812. https://doi.org/10.1177/0309132516660656

Arias M, Atienza M, Cademartori J (2014) Large mining enterprises and regional development in Chile: between the enclave and cluster. J Econ Geogr 14(1):73–95. https://doi.org/10.1093/jeg/lbt007

Aroca P, Atienza M (2011) Economic implications of long distance commuting in the Chilean mining industry. Resour Policy 36(3):196–203. https://doi.org/10.1016/j.resourpol.2011.03.004

Atienza M, Arias-Loyola M, Lufin M (2019) Building a case for regional local content policy: The hollowing out of mining regions in Chile. Extract Indus Soc 7(2):292–301. https://doi.org/10.1016/j.exis.2019.11.006

Bair J, Werner M (2011). Commodity chains and the uneven geographies of global capitalism: a disarticulations perspective. SAGE Publications Sage UK, London, England

Baldwin R (2019) The globotics upheaval. Globalization, robotics and the future of work. Oxford University Press, New York, United States of America

Barton J, Gwynne R, Murray W (2008) Transformations in resource peripheries: an analysis of the Chilean experience. Area 40(1):24–33. https://doi.org/10.1111/j.1475-4762.2008.00792.x

Breul M, Revilla Diez J (2018) An intermediate step to resource peripheries: the strategic coupling of gateway cities in the upstream oil and gas GPN. Geoforum 92:9–17. https://doi.org/10.1016/j.geoforum.2018.03.022

Bridge G (2008) Global production networks and the extractive sector: governing resource-based development. J Econ Geogr 8(3):389–419. https://doi.org/10.1093/jeg/lbn009

Castree N, Coe NM, Ward K, Samers M (2004) Spaces of work: global capitalism and the geographies of labour. Sage, London

Coe NM (2012) Geographies of production III: making space for labour. Prog Hum Geogr 37(2):271–284. https://doi.org/10.1177/0309132512441318

Coe NM, Yeung H (2015) Global production networks. Theorizing economic development in an interconnected world. Oxford University Press, New York

Coe NM, Hess M, Yeung HWC, Dicken P, Henderson J (2004) Globalizing' regional development: a global production networks perspective. Trans Inst Br Geogr 29(4):468–484. https://doi.org/10.1111/j.0020-2754.2004.00142.x

Garcés E (2003) Las ciudades del cobre. Del campamento de montaña al hotel minero como variaciones de la company town. EURE 29:131–148

Hayter R, Barnes T, Bradshaw M (2003) Relocating resource peripheries to the core of economic geography's theorizing: rationale and agenda. Area 35(1):15–23. https://doi.org/10.1111/1475-4762.00106

Henderson J, Dicken P, Hess M, Coe N, Yeung HWC (2002) Global production networks and the analysis of economic development. Rev Int Polit Econ 9(3):436–464. https://doi.org/10.1080/09692290210150842

Hobart CW (1979) Commuting work in the Canadian north: some effects on native people. Proceedings. Conference on commuting and northern development. University of Saskatchewan, Institute of Northern Studies, Saskatoon, February, 1–38

Irarrázaval F (2021) Natural gas production networks: resource making and interfirm dynamics in Peru and Bolivia. Ann Am Assoc Geogr 111(2):540–558. https://doi.org/10.1080/24694452.2020.1773231

Isaksen A (2015) Industrial development in thin regions: trapped in path extension? J Econ Geogr 15(3):585–600. https://doi.org/10.1093/jeg/lbu026

Jessop B, Brenner N, Jones M (2008) Theorizing sociospatial relations. Environ Planning D: Soc Space 26(3):389–401. https://doi.org/10.1068/d9107

Manky O (2016) From towns to hotels: changes in mining accommodation regimes and their effects on labour union strategies. Br J Indus Relat 55(2):295–320. https://doi.org/10.1111/bjir.12202
Manky O (2018) Resource mobilisation and precarious workers' organisations: an analysis of the Chilean subcontracted mineworkers' unions. Work Employ Soc 32(3):581–598. https://doi.org/10.1177/0950017017751820
Martinus K, Suzuki J, Bossaghzadeh S (2020) Agglomeration economies, interregional commuting and innovation in the peripheries. Reg Stud 54(6):776–788. https://doi.org/10.1080/00343404.2019.1641592
Massey D (1984) Spatial divisions of labour: social structures and the geography of production. MacMillan, London, UK
McElroy CA (2018) Reconceptualizing resource peripheries. The New Oxford Handbook of Economic Geography, July, pp 715–731. https://doi.org/10.1093/oxfordhb/9780198755609.013.32
McGrath S (2018) Dis/articulations and the interrogation of development in GPN research. Prog Hum Geogr 42(4):509–528. https://doi.org/10.1177/0309132517700981
Palomino J, Sarrias M (2019) The monetary subjective health evaluation for commuting long distances in Chile: a latent class analysis. Papers Reg Sci 98(3):1397–1417. https://doi.org/10.1111/pirs.12416
Paredes D, Fleming-Muñoz D (2021) Automation and robotics in mining: jobs, income and inequality implications. Extract Indus Soc 8(1):189–193. https://doi.org/10.1016/j.exis.2021.01.004
Paredes D, Soto J, Fleming D (2018) Wage compensation for fly-in/fly-out and drive-in/drive-out commuters. Papers Reg Sci 97(4):1337–1353. https://doi.org/10.1111/pirs.12296
Pérez D, Medel R, Velásquez D (2017) Radiografía de las huelgas laborales en el Chile del neoliberalismo democrático (1990–2015): masividad del conflicto por fuera de la ley en un sindicalismo desbalanceado. In: Ponce J, Santibáñez C, Pinto J (eds) Trabajadoras y trabajadores procesos y acción sindical en el neoliberalismo chileno (1979–2017). América en Movimiento, Santiago, pp 155–176
Phelps N, Atienza M, Arias M (2015) Encore for the enclave: the changing nature of the industry enclave with illustrations from the mining industry in Chile. Econ Geogr 91(2):119–146. https://doi.org/10.1111/ecge.12086
Phelps N, Atienza M, Arias M (2018) An invitation to the dark side of economic geography. Environ Planning A 50(1):236–244. https://doi.org/10.1177/0308518X17739007
Prada-Trigo J, Barra-Vieira P, Aravena-Solís N (2021) Long-distance commuting and real estate investment linked to mining: the case study of Concepción metropolitan area (Chile). Resour Policy 70:101973. https://doi.org/10.1016/j.resourpol.2020.101973
Reinert, H. (2018). Notes from a projected sacrifice zone. ACME: Int J Crit Geogr 17(2):597–617
Richalet JP, Donoso MV, Jiménez D, Antezana AM, Hudson C, Cortès G, León A (2002) Chilean miners commuting from sea level to 4500 m.: a prospective study. High Altitude Med Biol 3(2):159–166
Scholvin S, Breul M, Revilla Diez J (2020) A magic formula for economic development? Glob Market Integr Spatial Polar Extract Indus Area Develop Policy. https://doi.org/10.1080/23792949.2020.1823237
Sheppard E (2013) Thinking through the Pilbara. Aust Geogr 44(3):265–282. https://doi.org/10.1080/00049182.2013.817035
Storey K (2010) Fly-in/fly-out: implications for community sustainability. Sustainability 2(5):1161–1181
Storey K, Shrimpton M (1988) Long-distance commuting in the Canadian mining industry. Centre for Resource Studies, Queen's University, Kingston, Ontario
Torkington AM, Larkins S, Gupta TS (2011) The psychosocial impacts of fly-in fly-out and drive-in drive-out mining on mining employees: a qualitative study. Aust J Rural Health 19(3):135–141

Vodden K, Hall H (2016) Long distance commuting in the mining and oil and gas sectors: implications for rural regions. Extract Indus Soc 3(3):577–583. https://doi.org/10.1016/j.exis.2016.07.001

Miguel Atienza is Ph.D. in Economics from Universidad Autónoma de Madrid (Spain) and MPhil in Development Studies at the Institute of Development Studies (IDS) at the University of Sussex (UK). He is Professor of regional and urban economics at the Economics and Business Faculty of the Universidad Católica del Norte in Antofagasta (Chile). His main areas of research are regional and local development, mining and regional development, entrepreneurship and labour interregional mobility.

Printed by Books on Demand, Germany